FLORICULTURIST
Open Cultivation

THE AUTHOR

Dr. D. R. Singh is currently working as the Vice Chancellor, Chandrashekhar Azad Agricultural and Technological University, Kanpur. Born in 1964, his work experience is more than 28 years. During this time he has worked as Director of the Institute of Horticulture, Professor and Indian Council of Agricultural Research, New Delhi. Apart from this, various schemes have been embodied, guided, coordinated and implemented by him. He has been honored with 16 awards by various national and international organizations, including ICAR Fakhruddin Ali Ahmed, Lieutenant Governor's Commendation, Dr RS Paroda Award, Outstanding Scientist Award, Limca Book of National Records and Fellow awards etc. are prominent ones. Dr. Singh has developed 17 species of different horticultural crops as well as agricultural techniques of different crops and has successfully completed 14 projects funded by various institutions as principal investigator. He has published 206 research papers, 13 books, 50 book chapters and training bulletins at national and international level, as well as he is the chairman, expert and member of committees and scientific committees formed for various policy decisions. He is a renowned scientist in the field of horticulture at national and international level.

FLORICULTURIST
Open Cultivation

Prof. D. R. SINGH

2022

Associated Publishing Company

A Division of

Astral International Pvt. Ltd.

New Delhi – 110 002

ISBN: 9789354614675

Publisher's Note:

Every possible effort has been made to ensure that the information contained in this book is accurate at the time of going to press, and the publisher and author cannot accept responsibility for any errors or omissions, however caused. No responsibility for loss or damage occasioned to any person acting, or refraining from action, as a result of the material in this publication can be accepted by the editor, the publisher or the author. The Publisher is not associated with any product or vendor mentioned in the book. The contents of this work are intended to further general scientific research, understanding and discussion only. Readers should consult with a specialist where appropriate.

Every effort has been made to trace the owners of copyright material used in this book, if any. The author and the publisher will be grateful for any omission brought to their notice for acknowledgment in the future editions of the book.

Published by : **Associated Publishing Company**
A Division of
Astral International Pvt. Ltd.
– ISO 9001:2015 Certified Company –
4736/23, Ansari Road, Darya Ganj,
New Delhi-110 002
Ph. 011-43549197, 23278134
E-mail: info@astralint.com
Website: www.astralint.com

Preface

Floriculturist- Open Cultivation is a person who undertakes various activities of flower cultivation in open field condition, which involves preparation of the field, seed bed, planting, transplanting, selection of flower crops, and caring of flower crops. Floriculturist performs maintenance operations such as thinning, weeding, irrigation, fertigation, controlling pests and diseases, etc. The job is to be performed in an efficient manner to allow the production of high quality flowers, their harvesting and post–harvest management towards getting higher return.

First chapter Preparatory Flower Cultivation covers importance of floriculture and floriculturist, classification of flower crops, nursery raising, preparation of land for nursery, media for propagation of plants in nursery beds, soil testing, nutrient managements, IPM in the nursery, basics of the flower propagation, nursery tools and implements, use of growth regulators, selection and maintenance of mother trees, collection of scion wood stick ,techniques of propagation through specialized organs, corm, runners, suckers, plant protection, hi-tech nursery, high yield varieties of flowers, drought tolerance, transplanting, marketing survey and feasibility. In the second chapter, Crop Cultivation in Flower Crops have been covered with emphasis on types of flowers for cultivation, estimate nutrient requirement for flower crop, crop spacing and border crops, time of planting flower crop, seed rate, integrated nutrient management, method of application of fertilizers, specialized practices, irrigation management, pest and disease management. In the third chapter, Harvest and Postharvest Management in Floriculture deals with practicing flower harvesting, identify harvesting stage, handling harvested flowers, grading of different flower, use of chemicals for enhancing vase life of flower, value addition of flowers , packing of flowers, tools for flower harvesting, storage and transportation. Chapter four

Maintain Health and Safety at Work Place deals with health hazards at work place, practice general safety and first aid, maintain clean and efficient workplace, render appropriate emergency procedure, fire and emergency procedures.

This book has been designed to meet the needs of the floriculturists, students of Agriculture, Horticulture, Botany and layman flower enthusiasts, landscape professionals, horticulturists, nurserymen, amateur growers, enthusiasts, city and industry planners teachers as well as garden and nature lovers.

I am grateful to all those persons as well as various books, manuals, periodicals, magazines, journals etc. that helped in the preparation of this book. In spite of the best efforts, it is possible that some errors may have occurred into the compilation and editing of the book. Further queries, constructive suggestions and criticisms for the improvement of the book are always welcome and shall be thankfully acknowledged.

Author

Contents

Chapter 1

Preparatory Flower Cultivation

1.1 Importance of Floriculture and Floriculturist

1.1.1 Floriculture

Floriculture deals with the growing or cultivation of beautiful flowering plants. It is a branch of horticulture that deals with flower cultivation on a commercial scale, its marketing and selling. If taken on a broad perspective, it includes even a flower arrangement. The common plants grown for their flowers are different varieties of Chrysanthemum, Aster, Dahlia, and other Composite plants, Gladiolii and some flowers of Liliaceae and Amarilladaceae, Orchid varieties, Rose varieties, *etc.*

Importance

Flowers symbolize purity, beauty, peace, love, and passion. To a Japanese flower-arranger, each flower expresses one or more meaning while to the Indian, flowers have much greater significance. The aesthetic value of flowers in our daily life cannot be over-emphasized. In our society, no social function is complete without the use of flowers. Floral garland, gajaras and venis are needed for marriage ceremonies. Floral ornaments, bouquets, or flower arrangements also find a place in social gatherings, birthday parties, welcoming home-coming friends or relatives, and honoring dignitaries. The arrival of new-born has rejoiced with flowers, the sick are wished speedy recovery by offering flowers, while the dead people are bidden farewell with flowers along with tears of sorrow.

Goals

- ☆ Crop improvement - Genetic resource enhancement, evaluation, and conservation.

- Breeding of novel color, short duration, temperature, and drought-tolerant ornamental cultivars.
- Breeding of dwarf cultivars of high-value flowers/foliage plants.
- Strengthening/standardization/popularization of F1 hybrid seed production in important annual flowers.
- Exportable 'made in India' varieties using molecular breeding and advanced techniques.
- Production technology.
- Development of agro-technologies for open field and protective cultivation – region-specific, energy-efficient, low cost and eco-friendly production systems.
- Commercialization of 'Specialty Flowers' as new flowers to attract buyers and market demand.
- Digitalized spray and fertigation schedules.
- Production of quality planting materials through tissue culture.
- Improved media, new cladding materials, and alternate energy and light sources.
- New generation molecules for enhancing blooming span, shelf life, *etc.*
- Mechanization of planting to packaging and storage.
- Postharvest technology: Standardization of operating procedures for isolation of pigments, essential oils, and natural colors/dyes.
- New flower crops are suitable for making dry flowers and technologies for efficient drying.
- Natural dyes for pot parties and other flower arrangements.
- Postharvest engineering for improved packaging and low-cost storage to reduce perishability.
- Research on anti-senescence technology to delay senescence.
- Landscaping - Improved grass species for turfs.
- Urban greening technologies and vertical gardening.
- New plants are suitable for landscaping.
- Plant health management.
- Healthy landscape and beautiful flowers.
- An inventory of pests and pathogens of flower crops.
- Digitalized pest and disease forecasting expert system.
- Prevention/management/eradication of the emerging pests and pathogens under changing climatic scenarios.

- ☆ Crop protection strategies with reduced pesticide inputs.
- ☆ Multiple pests and disease resistant flowering plants.
- ☆ Assured plant Biosecurity by generating database on potential pests and diseases of flower crops, diagnostics and containment techniques and information sharing.
- ☆ Quality floriculture products meeting international sanitary and phytosanitary standards.

Floriculture in India

Floriculture is an age-old farming activity in India having immense potential for generating gainful self-employment among small and marginal farmers. In recent years it has emerged as a profitable agri-business in India and worldwide as improved standards of living and growing consciousness among the citizens across the globe to live in the environment-friendly atmosphere has led to an increase in the demand of floriculture products in the developed as well as in the developing countries worldwide. The production and trade of floriculture have consistently increased over the last 10 years. In India, the Floriculture industry comprises flower trade, production of nursery plants and potted plants, seed and bulb production, micropropagation and extraction of essential oils. Though the annual domestic demand for the flowers is growing at a rate of over 25 per cent and international demand at around Rs 90,000 crore India's share in the international market of flowers is negligible. However, India is having a better scope in the future as there is a shift in the trend towards tropical flowers and this can be gainfully exploited by countries like India with a high amount of diversity in indigenous flora.

After liberalization, the government identified floriculture as a sunrise industry. The liberalization of industrial and trade policies paved the way for the development of export-oriented production of cut flowers. The new seed policy has already made it feasible to import planting material of international varieties. Floriculture products mainly consist of cut flowers, pot plants, cut foliage, seeds bulbs, tubers, rooted cuttings, and dried flowers or leaves. The important floricultural crops in the international cut flower trade are rose, carnation, chrysanthemum, gerbera, gladiolus, orchids, anthurium, tulip and lilies.

Scope of Floriculture in India

In India, floriculture is only a developing subject that offers much scope for improvement. This problem can be tackled from several angles such as conservation, domestication, and introduction. India is rich in its plant resources, many of which are of ornamental value and some are potentially ornamental. Much of this wealth is wasted as a result of rapid urbanization, industrialization, and unscrupulous collection. For example, some unscrupulous nurserymen are selling one beautiful Paphiopedilum insigne orchid plant for the price of a single flower in foreign markets.

The domestication of wild plants with potential ornamental value is another way of improving garden wealth. In the process of domestication, possibly in an altogether different climate, the wild plants generally pass through many changes, which itself may cause some chance of improvement.

The introduction of plants either from the wild with potential ornamental value or from abroad offers good scope for enriching ornamental plants. Initially, we should introduce wild plants with potential ornamental value, which can be improved upon by systematic breeding. Simultaneously, plants of proven quality can be introduced from abroad and acclimatized in our country. The process of introduction has improved our ornamental horticulture to a great extent. The introduction of Mussaenda philippica, the double-bracted mussaenda with white, pink, and red colors, from Thailand have acclimatized very well in the humid and mild climates of Kolkata, Bangaluru and Kerala. The introduction of many beautiful bougainvilleas from abroad, *e.g.* 'Lady Mary Baring', 'Golden Glow', and the multi-bracted cultivars such as 'Mahara', 'Cherry Blossom', *etc.* has enriched our ornamental horticulture to a great extent.

1.1.2 Floriculturists

Floriculture or flower farming deals with the study of the growing and marketing of flowers and foliage plants. It includes the cultivation of flowering and ornamental plants for direct sale or for use as raw materials in the cosmetic and perfume industry and in the pharmaceutical sector. It covers the production of planting materials by seeds, cuttings, budding, and grafting. Floriculture can be defined as the art and knowledge of growing flowers to perfection and the person associated with such fields is known as floriculturist.

The Floriculturist deals with the cultivation and growth of flowers and foliage plants. The ornamental plants and flowers are used mainly as a raw material in the pharmaceutical industry, in the perfume industry, cosmetic industry and so on.

The work of floriculturist includes:

- ☆ Selection of the right site for greenhouse construction
- ☆ Obtaining appropriate equipment and tools
- ☆ Testing of water and soil
- ☆ Procuring seed material
- ☆ Selecting and treating seed to be sowed
- ☆ Preparing of land and nursery
- ☆ Transplanting the seedling
- ☆ Cultivating the flower crop
- ☆ Harvesting the produce
- ☆ Marketing and selling the farm produce

The main expertise of floriculturist includes:

- ☆ Knowledge of tools and equipment to be used in flower cultivation
- ☆ Selecting an appropriate site for cultivation of flowers
- ☆ Proficiency in preparing the land for flower cultivation
- ☆ Ability to do propagation and transplantation of seedlings
- ☆ Knowledge about nutrient management, integrated insect, pest and disease management for plants
- ☆ Familiarization about irrigation management, harvest, and postharvest management
- ☆ Skilled in basic farm management
- ☆ Equipped to assimilate market information
- ☆ Ability to update with the latest development in a variety of flower seeds and new equipment

1.2 Classification of Flower Crops Based on different Criteria

Flower Crops are Classified Based on

Season of Growing

- ☆ Summer season Annuals: Zinnia, Kochia, Portulaca, Tithonia, Gaillardia, Gomphrena, Sunflower, Daisy, *etc.*
- ☆ Rainy season annuals: Balsam, cock's comb, Celosia, Gaillardia, *etc.*
- ☆ Winter season annuals: Antirrhinum, China aster, Cornflowers, Larkspur, Sweet Sultan, Phlox, Verbena, Candytuft, Petunia, *etc.*

Flower Colors

- ☆ White flowering: Antirrhinum. Alyssum, Dianthus, China aster, Zinnia, Chrysanthemum, Gladiolus, Gerbera, *etc.*
- ☆ Purple, Lavender or Blue: Daisy, Rose, Carnation, Dahlia, Ageratum, China aster, Clitoria, Delphinium, Petunia, Viola, Verbena, Tithonia, Torenia, Daisy, *etc.*
- ☆ Yellow or orange: Antirrhinum, Marigold Calendula, Zinnia, Gaillardia, Rose, Gladiolus, Carnation, *etc.*
- ☆ Red pink: Antirrhinum, Rose, Gladiolus, Carnation, Gerbera, Dahlia, *etc.*

Purpose of Growing

- ☆ Rockery: Ageratum, Alyssum, Brachycome, Phlox, Portulaca, Linum, Nemesia, Saponaria, Godetia, Euphorbia, *etc.*

- Hanging basket: Dwarf Ageratum, Petunia, Portulaca, Verbena, Torenia, Begonia.
- Edging of beds or paths: Dwarf Ageratum, Alyssum, Brachycome, Dianthus, Nigella, Portulaca, *etc.*
- Fragrant flowers: Sweet Alyssum, Sweet Sultan, Sweet pea, Stock, Phlox, Carnation, Rose, Jasmine, Tuberose, *etc.*
- Bedding purpose: Dahlia, Marigold, Phlox, Verbena, Carnation, Petunia, Ice Plant, Candy Tuft, Balsam, Portulaca, *etc.*
- Aromatic: Rose, Jasmine, Tuberose
- For pots: Carnation, Chrysanthemum, Dahlia, Rose, Antirrhinum, Petunia, Agloenema, Alocasia, Anthurium, Orchids, Aralia, Begonia, Chlorophytum, Dracena, *etc.*
- For loose flowers: Marigold, Jasmine, Crossandra, Berlaria, Chrysanthemums, China Asters, Sunflowers, Zinnia, Gaillardia, Rose, Dahlia, *etc.*
- For Dry flowers: Statice, Helichrysum, Acroclinum, Gomphrena, Limonium, Marigold, Rose, Lady's Lace, Nigella, *etc.*
- For Hedge purpose: Lawsonia, Duranta, Tecoma, Bougainvillea, Thevetia, Hibiscus, Murraya, Dodonea, Acalypha, Aralia, Ipatorium, Chlorodendron spp, *etc.*
- Cut flowers/modern flowers: Rose, Chrysanthemum, Carnation, Gerbera, Anthurium, Orchids, Gladiolus, Tuberose, *etc.*

Nature of Growth

- Annuals: Nasturtium, Ice plant, Hollyhock, Sweet pea, Annual Chrysanthemum, Carnation, Cornflower, Sweet Alyssum, Dahlia, Marigold, Verbena, Phlox, Pinks, Calendula, *etc.*
- Perennials: Rose, Jasmine, Crossandra, Anthurium, Orchids, Chrysanthemum, Berlaria, Hibiscus, Gerbera, Carnation, Bulbous Crops.

Mode of Propagation

- Bulbous plants: Lily, Narcissus, Tulip, Tuberose,
- Cormellous plants: Gladiolus, Crocus.
- Rhizomatous plants: Canna, Iris,
- Tuberiferous plants: Dahlia

Usage/Purpose

- Traditional flowers: They are offered in religious and social ceremonies, used as an adornment by women, and offered for worship at home as well as in temples.

e.g. Jasmine, Rose, Mums, Marigold, Crossandra, Tuberose, Berlaria, Gaillardia, *etc.*

- ☆ Non-traditional flowers: They are referred to as cut flowers or modern flowers. Generally, flowers are harvested along with a long stem/stalk.

  e.g. Hi-tech roses (Dutch roses), Gerbera, Carnation, Chrysanthemum, Gladiolus, Orchids, Anthuriums, Goldenrod, Bird of Paradise, Limonium, Liatries, Stock, *etc.*

- ☆ Industrial Flowers/Value Added flowers: They are used as a raw material in industries for extraction of essential oil and also preparation of some edible products like gulkand, gulroghan, pigments as natural colors and also dry flowers.

  e.g. Tuberose, Jasmine, Rose, Marigold, *etc.*

1.3 Nursery Raising

1.3.1 Nursery

Nursery is a place where plants are grown, take care of and sold out. It is also a place where plants, shrubs and ornamental trees are grown for transplanting into landscape areas.

Plants Grown in Nursery in Pot/Container

In Nursery, plants can be grown on fields or in a pot/container. Plants that are grown in pot/container are of shorter to medium height and can be easily movable.

1.3.2 Nursery Categories

Nurseries can be categorized as:

1) Temporary Nursery

- ☆ Such a type of nursery is developed only to fulfill the need of season or targeted project.
- ☆ Here, the production of seedlings of transplanted vegetables and flower crops are temporary in nature.
- ☆ In this, temporary arrangement of growing of forest seedlings for planting in an area is done.

Advantages

- ☆ It can be shifted to any site.
- ☆ It needs low capital cost.
- ☆ It reduces transportation cost.
- ☆ It works within climatic site conditions.
- ☆ In this, skilled labour is not required.
- ☆ It can be near the planting site.

Disadvantages

- ☆ In this, the water supply can be poor.
- ☆ Soil is also poor in this.
- ☆ There appears a shortage of trained labour.
- ☆ All can result in poor quality trees.

2) Permanent Nursery

- ☆ Such a nursery is placed permanently so as to produce plants regularly.
- ☆ The nursery has all permanent features and has permanent mother plants.
- ☆ In such type of nursery, the work can be done on a regular basis all the year-round.
- ☆ With the type of plants produced, nurseries can be classified as:
  - ❖ Fruit plant Nurseries: In this, seedlings and grafting of fruit crops are prepared.
  - ❖ Vegetable nurseries: In this, seedlings of cauliflower, cabbage, brinjal, and tomato are prepared.
  - ❖ Flowers plants nurseries: In this, seedlings of flowering plants like

gerbera, carnation, petunia, salvia, rose, chrysanthemum, coleus, aster, dianthus are developed.

- Forest nurseries: In this, seedlings of plants that are useful for forestation like pine, oak, teak, eucalyptus, casuarinas are prepared and sold.
- Different nurseries: In this, plants having high economic value, rare and medicinal, herbal plants are propagated such as geranium, rose, calendula, and marigold.

Advantages

- It is well designed.
- It has a good water supply.
- It provides good quality and required a number and size of plants.
- It has good soil.
- It can be well managed and protected.
- In this, planting material is always available

Disadvantages

- It requires high capital and transportation cost.
- It is not always in climatic site conditions.
- It is not always near the planting site.
- In this, permanent and skilled labor is required.

3) Dry Nursery

- It is a type of nursery that is maintained without any irrigation or other artificial watering.

4) Wet Nursery

- It is a type of nursery that is maintained by irrigation or other artificial watering during the dry periods.

1.3.3 Importance of Nursery

A nursery is a part of agriculture where plants are propagated, nurtured, grown, and sold out for the home garden or commercial purpose. The main purpose of nurseries is to supply young plants or saplings to gardens, farms, agriculture, forestry, and conservation. Nurseries can have stock of various plants such as herbs, shrubs, plants, small trees, big trees, and bonsai. A nursery plant center offers to its customers such as home gardeners, landscape gardeners or commercial farmers, all types of young plants that are native and imported.

Establishment of the nursery is important for the production of:

- Proposed size and age of seedlings.

- ☆ Desired number of seedlings.
- ☆ Healthy plants of the desired spp.
- ☆ Plants of superior quality.
- ☆ Desired plants at the right time.
- ☆ Seedlings at a cheaper cost.

Importance

- ☆ In the nursery, seedlings and grafts are produced from which the fruit orchards and ornamental gardens can be established with minimum care, cost, and maintenance.
- ☆ The nursery planting materials are available at the start of planting the season which helps in saving time, money and efforts of farmers to raise seedlings.
- ☆ There is a wide scope for fruit orchards, ornamental, vegetable, and landscape gardens at public places, highways, and co-operative housing societies.
- ☆ It assures the production of genetically improved quality planting material.
- ☆ It provides employment opportunities for technical, skilled, semi-skilled, unskilled labor.
- ☆ They are important sources for the supply of seedlings for fruit, pulp and paper, fuelwood, timber, and other industrial demands.

1.3.4 Nursery Raising

The nursery raising is an important step in flowers. This method is the easiest and cheapest which is normally employed on the commercial sale. Raising nursery in flower crops is very important as almost all flowers are first raised in the nursery and then are transplanted in the field at the proper spacing.

Advantages

1) It is convenient to look after baby seedlings.
2) It is possible to provide favorable growth conditions *i.e.* germination and growth.
3) Eliminate the problem of difficult soils.
4) Easy weed control
5) Reduced field management costs
6) Improved crop uniformity
7) Higher yields

8) More optimal use of Hybrid seeds
9) Shorter, growing season and more efficient use of land
10) More accurate prediction of harvest date.

Disadvantages

1) Transplant shock which delays growth but is not as severe on the cell raised seedlings compared to bare-rooted seedlings.
2) Cost of seedlings which adds to production cost.
3) Extra labor to establish the crop.

1.3.5 Factors Raising a Nursery

- ☆ Location of the nursery
- ☆ Near the house
- ☆ Well exposed to the sun but protected against severe heat
- ☆ Well protected against animal damage, strong winds.

Factors to be considered for raising a nursery will include:

Water

- ☆ Near water source
- ☆ Continuous supply of good water
- ☆ Seedbed to be kept moist not continually wet.

Soil

- ☆ Soil to have a large quantity of organic matter
- ☆ Soil texture should be neither too coarse nor too fine.
- ☆ Fair degree of water holding capacity
- ☆ Rich in necessary elements

1.3.6 Procedure for Raising Nursery

- ☆ Select a suitable site for preparation of raised beds;
- ☆ Dig soil at least 45 cm deep and well-decomposed compost/farmyard manure and evenly spread;
- ☆ For good irrigation, each bed is subdivided into 3-4 parts, and the narrow trench is dug on one side;
- ☆ For winter season flower seeds to be sown in June-July while for the spring season, seeds to be sown in November - December;
- ☆ Irrigate the nursery with water;

- ☆ Protect the nursery against the hot sun and heavy rains;
- ☆ Remove the unwanted plants regularly.

1.4 Site Selection Criteria

The selection of the area for the nursery is critical, and it should be sited as centrally as possible to the field(s) to be planted. Nurseries can be started irrespectively of where a florist is located but with enough land space to grow plants. Nursery location may be predetermined based on cost investment, availability of the property. The choice of Nursery location will directly influence marketing the plant saplings, manpower, and future expansion. The location of the nursery must be near to major cities and townships, transportation, availability of labor and convenience for future expansion. Nursery site location must be nutrient-rich or medium soil with a good water source. Nursery site must have good soil drainage and free from soil pathogens.

1.4.1 Factor to be Considered while Selecting the Nursery Site

There are certain factors which need to be considered while selecting a nursery site such as:

Topography

- ☆ The selected area should be flat with gentle slopes between zero and 30.
- ☆ The area should have a reliable/permanent source of water supply for irrigation.

Water Supply

- ☆ Water requirements as quality and quantity to be determined before starting of site preparation.
- ☆ The easiest source of water is where there is a large natural pond or lake whereby all that is required is to place an intake pipe to connect with a pump unit.
- ☆ A back-up system should be considered, mainly in isolated areas or areas of lower or unreliable rainfall.

Drainage

- ☆ The site selected should not be prone to flooding, as it will damage the seedlings and buildings/stores.

Area/Spacing

- ☆ For good growing conditions with minimal risk of etiolating, nursery should have planting density of 13,800 polybags/ha with 0.91m(3ft) x 0.91m(3ft).

- ☆ The spacing should be increased by 0.15m (0.5ft) if seedlings are anticipated to be kept in the nursery for more than 12 months.

Accessibility/Approach

- ☆ Walk passage in nursery and their alignments should be carefully planned and laid as per the placement distance of polybags and type of irrigation to be used.

- ☆ Approach to nursery should be sufficiently wide to allow vehicles to move during peak planting periods to facilitate supervision and movement of materials.

1.5 Preparation of Land for Nursery

The preparation of area for nursery is important to allow actual seedling growth, maintenance of nursery site, unimpeded access and to provide hygienic conditions for plant growth. In site preparation, there are four main activities involved such as:

- ☆ Design,
- ☆ Clearing,

- ☆ Fencing
- ☆ Lining

1.5.1 Nursery Design

A well-designed nursery allows access of many vehicles during evacuation of seedlings for field planting especially for large-scale plantings. The design of site can be done by drawing up of plan to show all paths, roads and irrigation points as shown.

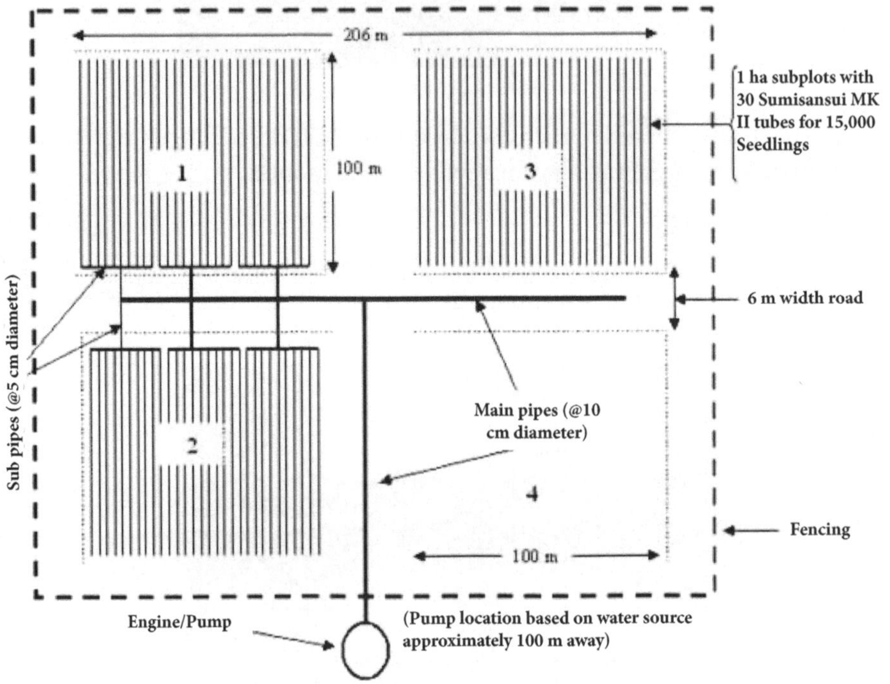

1.5.2 Nursery Clearing

- ☆ With boundaries determined, felling and clearing should be done at least 2 months before the arrival of seeds.
- ☆ After the clearance, do the fencing of the area, fill the polybags and then install the irrigation system.

1.5.3 Nursery Fencing

The main types of fences used for nurseries involves:

Conventional Barbed Wire Fence

- ☆ The specifications for conventional fence depend on type of animals from which, nursery needs to be safe guard.

☆ In case of four-strand barbed wire fence, the wires are spaced at 0.3, 0.6, 0.9 and 1.2 m from ground which will not allow dogs, cat, cattle and goats to enter inside the nursery.

Electric Fence

☆ Electric fence is used where there is a threat of wild mammalian pests.

☆ In this, wires of 250-300 kg breaking strain are used.

☆ The heights at which the electrified wires are placed above ground level depends on mammalian pest's height.

Lining

☆ Lining is done to give space to polybags evenly so that seedlings have good and uniform access to sunlight and can get good irrigation system.

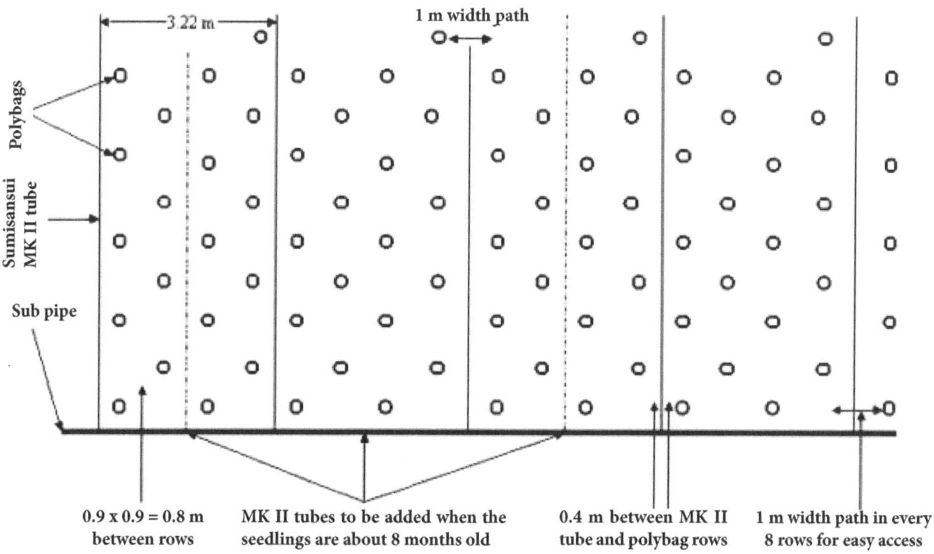

- ☆ Polybags are lined at 0.91m (3 ft) x 0.91m (3 ft) triangular spacing to give each seedling, best growth space.
- ☆ All seedling rows must be straight along the axis at 60° to each other and parallel to irrigation lines.
- ☆ In Sumisansui irrigation system, MK II tubes are laid down between polybags at every 4 rows, a 1 m wide path is provided at every 8 rows for easy access.
- ☆ When seedlings are about 7–8 months old, additional MK II tubes are added between the existing MK II tubes to give final layout of 1 tube for every 2-seedling row as shown.

1.6 Media for Propagation of Plants in Nursery Beds

1.6.1 Plant Propagation Method

The methods of propagation of plants can be classified as follows:

1. **Native Propagation**
   - ☆ It needs the participation of male and female germ cells form subsequent formation of the seeds.
   - ☆ Plants that grow from seeds have combined characteristics of both plant-parents.
2. **Asexual Propagation**
   - ☆ In this, plants are proliferate by using the specific parts without mediation of fertilization.
   - ☆ Plants that are obtained are similar to the parent plants.
   - ☆ They are divided into natural and artificial.

Natural Asexual Propagation

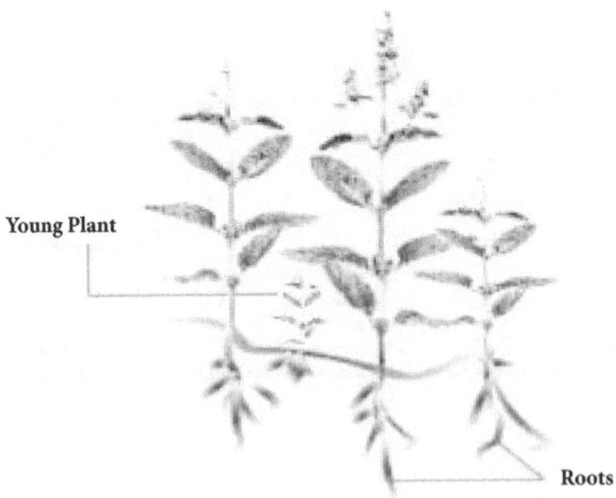

Suckers

- These are shoots that grow at the base of the plant at the existing root system and gradually develop their own independent root system.
- They are severed from parent plant together with their roots and cultured as individual plants.

Stolon

- These are modified shoots that grows horizontally on soil surface or below it, and have ability to form new independent plants or at their end points or at points where their eyes exist.

Bulbs, Tubers, Corms, Rhizomes

- These are modified and reproductive parts of root system of specific plant species.

Scion Plants

- They are small plants that fully grow up in mother plant that gradually develops their own root system and can develop autonomously.
- It is seen in species of Kalanchoe, Tillandsia, genre and certain types of orchids and aquatic plants.

Artificial Asexual Propagation

Cuttings

- These are parts of plant which under certain conditions can form a root system and grow into new plants.

Division

- It shows plants that develop tubers, bulbs, stems and rhizomes.

1.6.2 Pots/Container for Propagation and Growing Young Plants

Earthen Pots

- These are made of burnt porous clay in different sizes to provide the required amount of soil and root space to different kinds and sizes of plants.
- These have straight sides and are made wider at top than at the bottom to hold maximum bulk of compost where the feeding roots are and also to facilitate easy removal of soil, intact with roots at the time of planting or repotting.

- Gardener should consider tube pots of different sizes such as:

Tube Pot Sizes	*Height (cm)*	*Diameter (cm)*	*Cost per Pot (Rs.)*
Tube pot	20	13	15.00
¼ size pot	18	22	15.00
½ size pot	20	27	30.00
¾ size pot	25	32	50.00
Full size pot	35	35	65.00
Tub size pots	35	50	90.00

Seed Pan/Boxes

- These are the shallow earthen pots of about 10 cm high and 35 cm in diameter at the top.
- These have one large hole for drainage in the center or 3 holes at equidistant from each other.
- These boxes are made of wood, 40 cm wide and 60 cm long and 10 cm deep, with 6-8 properly spaced holes drilled in the bottom.
- Against each hole, a crock with its concave side down are put over it and also by the side of this crock, some coarse sand 2 or 3 handfuls are sprinkled on crock pieces which forms thin layer to prevent fine soil from clogging the drainage.
- Delicate seeds like Cineraria, Begonia, *etc.* are best sown in these containers.

Polythene Bags

- Small polythene bags with holes punched in the bottom for drainage and filled with porous rooting medium are used for propagation of cuttings like Jasmines, Durant, and Crotons *etc.* in the mist chamber.
- The young seedlings which are raised in nursery are subsequently transplanted in polythene bags and are kept there till they attain required growth for transplanting them to main field.

Plastic Pots

- The plastic pots, round and square are used to keep indoor plants.
- They are reusable, light weight, non-porous and requires only little storage space

1.6.3 Growing Media in Nursery

Functions of Growing Media

A growing medium serves four functions:

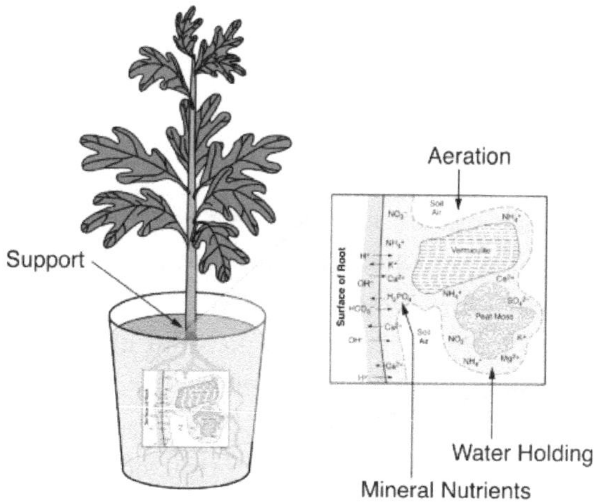

1. ***Physical Support***

   ☆ The growing medium must be porous but should provide physical support.

   ☆ Young plants are fragile which should remain upright so that they can photosynthesize and grow properly.

   ☆ With larger nursery items in single containers, the growing medium must be heavy enough to hold the plant upright against the wind.

2. ***Aeration***

   ☆ Plant roots need a steady supply of oxygen to convert the photosynthate from the leaves into energy so that the roots can grow and take up water and mineral nutrients.

   ☆ The byproduct of such respiration is carbon dioxide which should be dispersed into the atmosphere to prevent the buildup of toxic concentrations in the root zone.

   ☆ The gas exchange from this takes place in large pores or air spaces in a growing medium as nursery plants grow rapidly, which needs a medium with good porosity.

3. ***Water Supply***

   ☆ Nursery plants use a large amount of water for growth and development, so water supply should be provided by growing medium.

- Growing media are formulated so that they can hold water in small pores between their particles.
- Many growing media has a high percentage of organic matter like peat moss and compost as these materials have internal spaces that can hold water like a sponge.
- The growing media should have adequate porosity to absorb and store large amounts of water needed by the growing plants.

4. ***Supply of Mineral Nutrients***
   - Many essential mineral nutrients that nursery plants needed for rapid growth should be obtained by roots from growing medium.
   - Many mineral nutrients are electrically charged ions while positively charged ions (cations) include ammonium nitrogen (NH_4^+), potassium (K^+), calcium (Ca^{+2}), and magnesium (Mg^{+2}).
   - The cations are attracted to negatively charged sites on growing medium particles up to the point when roots extracted cations.

Development of Growing Medium

It is seen that as many recipes exist for growing media as a number of nurseries, there are no specific global recommendations. Gardeners should always involve in experimenting and finding suitable, local, affordable ingredients so as to create, good growing media. There are generally three types of growing media used in pot/ container nurseries such as:

1. ***Seed Propagation Media***

For germinating seeds or establishing germinant, the medium should be sterile having fine texture so as to maintain high moisture around the germinating seeds.

2. ***Rooting Cuttings Media***

Cuttings are rooted in frequent misting, so the growing medium should be very porous to prevent water logging and allowing good aeration required for root formation.

3. ***Transplant Media***

When smaller seedlings or rooted cuttings are transplanted into large containers, the growing medium needs to be coarser.

Preparation of Nursery Beds

Nursery bed is a place of high land for raising Seedling and is made with pillar bricks and furrows, *etc.* which is like a seed bed. This place is selected for the planting of cutting, layers, and grafts instead of seed sowing. A poorly constructed and managed nursery bed will lead to poor germination or diseased seedlings. The seedlings may fail to grow when transplanted.

A nursery bed is a prepared plot of land for raising seedlings. It acts as a temporary home for young plants until they are eventually planted in the main garden. The nursery bed should be protected from strong winds and should never be located on a slope to avoid erosion of seeds and nutrients, it should face east-west direction and should never be put in a water logged area.

For raising the seedlings two types of nursery bed are prepared in the nursery.

- ☆ Bed for direct raising of Seedling.
- ☆ Bed for rising of seedlings in the poly bags.

1. *Bed for Direct Raising of Seedling*

- ☆ Initially the land should be prepared for making a bed, so for this, land is to be ploughed and grasses arc to be rooted out where debris should be removed down the clods.
- ☆ If the soil is bad for making nursery bed, then fertile soil should be brought out from other location that should be mixed thoroughly.
- ☆ If 1 cubic meter (m^3) cow dung or compost is required for each 10 square meters of nursery bed and 1.5 kg of urea, TSP and MP at ratio of 4:2:3 are mixed with soil of bed, then surface of land needs to be levelled.
- ☆ The bed should be height of 10 to 15 cm by making furrows/drain in and around the bed, so edge of bed is to be bounded.
- ☆ Post of wood or bamboo with height 30 to 35 cm is to be settled down vertically to the Soil around the bed, so branches of bamboos to be attached with the post.
- ☆ The edge of bed can also be made with bricks.

2. *Bed of Rising of Seedling in Poly Bag*

- ☆ The terms of poly bag, the land is not to be ploughed.
- ☆ The surface layer is to be levelled by earthing up of soil in the middle portion of two beds with a height of 10 to 15 cm.
- ☆ The edge is made up of bed where edge of bed can be made with bricks or bamboo.

Sowing of Seeds

Sowing seeds is another step in crop production which can be done once soil preparation is done. Sowing is a process of planting the seeds in the soil which are sown that needs careful selection for high quality yield.

Methods of Sowing Seeds

There are various methods of sowing the seeds such as:

Sowing by Hand

- ☆ The scattering of seeds by hand is a method of sowing seeds.
- ☆ It is also known as broadcasting method.
- ☆ It is an economical method which is mostly used.
- ☆ In this, distribution of seeds is not uniform which result in clustering of seeds on the field.

Seed Drill

- ☆ It is a modern method of sowing seeds.
- ☆ It is a better and more efficient method than sowing by hand.
- ☆ It is usually done by attaching iron drills to a tractor.
- ☆ Seed drills ensure that the seeds are planted at equal intervals and at the correct depth in the soil.

Precautions to be taken while Sowing Seeds

Sowing seeds is essentially the most important part of crop production. It is necessary that the gardener should focus on:

- ☆ Seeds must be planted at the correct distance or intervals from each other which ensures that all plants get their fair share of light, water, and nutrients for growth and development. Planting seeds at equidistant has been proved to increase the yield of the farm.
- ☆ Seeds must be sown at the correct depth in the soil. If seeds are simply scattered on the top, they are likely to be blown away or eaten by animals or birds. If gardeners sow seeds too deep in-ground, they will not germinate due to lack of air.
- ☆ The seeds that a gardener sow should be of the highest quality which should be free from germs and diseases.

1.7 Soil Testing

1.7.1 Soil Test

Soil test refers to one or more of a wide variety of soil analyses which is conducted for one of several possible reasons. A soil test shows the analysis of a soil sample that determines nutrient content, composition, and other characteristics like acidity or pH level. A soil test determines fertility, or expected growth potential of the soil which shows nutrient deficiencies, potential toxicities from excessive fertility and inhibitions from the presence of non-essential trace minerals.

Soil testing is performed by commercial labs that offer a variety of tests, targeting groups of compounds and minerals. It involves collecting soil samples, preparation for analysis, chemical or physical analysis, interpretation of analysis results, and finally making fertilizer and lime recommendations for crops. The laboratory tests often check for plant nutrients in three categories such as:

- ☆ Major nutrients: nitrogen (N), phosphorus (P), and potassium (K)

- Secondary nutrients: sulfur, calcium, magnesium
- Minor nutrients: iron, manganese, copper, zinc, boron, molybdenum, chlorine

1.7.2 Importance

Soil testing is an important diagnostic tool that determines the nutrient needs of plants and environmental assessments. It facilitates fertilizer composition and dosage selection for land employed in both agricultural and horticultural industries. Soil testing can also be used to identify application rates of waste materials containing nutrients or other elements that could harm the environment.

Soil testing is the best method available for determining how much nutrient is needed for the optimum production. For certain crops, the use of plant analysis serves as a preferred method of determining crop nutrient needs. Making nutrient-application decisions without soil testing or plant analysis can be a risky endeavor because this will probably result in reduced profitability from either under or over-applying nutrients; in addition, over-application of nutrients can result in reduced water quality. It is important to follow the recommended procedures for taking soil and plant samples.

1.8 Nutrient Managements

1.8.1 Introduction

All plants require nutrients to survive and grow. They take nutrients from the air, soil, and water as you cannot see nutrients since they are colorless gases or like dust that is dissolved in water or stuck on each piece of soil.

The quantity of nutrients available to nursery plants is affected by:

- Substrate quality
- Water quality
- Plant type.

1.8.2 Features

- Nutrients are taken up by fine root hairs, not by the big roots.
- Even very large trees have many small, fine root hairs to absorb the nutrients and water they need.
- The larger roots are used for supporting the tree and for the storage of water and other plant food.
- The root hairs excrete liquids that affect the acidity of the soil (pH) and when pH changes, the number of nutrients present also changes.
- There are many plant food nutrients that are essential for proper seed development.

- Each of these nutrients is equally important to the plant, but each required in large quantities.
- The primary (macro) nutrients such as nitrogen, phosphorus, and potassium are most frequently required in the nursery.
- The deficiency due to the primary nutrients will be expressed by seedlings through deficiency symptoms, so the required fertilizer needs to be applied as required.
- In case of general nutrient support, FYM and compost are mostly used in nurseries where available soils are either too sandy or too heavy with low quality as it will help in building good soil structure, improves water holding capacity of the soil and provides nutrients for plants.
- Such nutrients greatly reduce the need for chemical fertilizers and if mixed with small amounts, dilute the fertilizer, making it available in much larger useful quantities.
- The gardener should therefore economically use required chemical fertilizers and natural fertilizer.
- The leaves of seedlings planted with unfinished compost usually turn yellow as the plant cannot get all nutrients as it needs whilst the immature compost that continues to absorb what little nitrogen is available.
- Adding decomposed manure in nursery mixture will assure of production quality and healthy seedlings.
- To boost rhizome growth, phosphorous should be added while the application of urea will result in foe good vegetative/foliar growth of seedlings.

1.8.3 Types of Nutrients

There are two types of nutrients:

- Macronutrients: required in large quantities
- Micronutrients: required in small quantities

macronutrients	micronutrients
Nitrogen (N)	Iron (Fe)
Phosphorus (P)	Manganese (Mn)
Potassium (K)	Zinc (Zn)
Calcium (Ca)	Copper (Cu)
Magnesium (Mg)	Boron (B)
Sulphur (S)	Chlorine (Cl)
	Molybdenum (Mo)

Nitrogen, phosphorus and potassium, together comprise over 75 per cent of mineral nutrients found in plant.

Nitrogen (N)

- ✰ It has similar benefits to protein.
- ✰ It is important for good stalk growth by excess decreases immunity and fruit ability.
- ✰ Nitrogen recommendations for many crops are based on yield goals for those crops.
- ✰ It is important to establish realistic yield goals for each field based upon historical yield data.
- ✰ Recommended application rates for nitrogen should not be exceeded.
- ✰ The use of the Pre-Sidedress Nitrogen Test is recommended in early summer after forage legumes or manure and biosolids applications to corn in order to show if extra nitrogen is needed.
- ✰ Residual values for nitrogen from legumes in rotation or previous applications of manure or sludge are deducted from total nitrogen recommendations.
- ✰ Growing winter cover crops is an effective practice for reducing nitrate losses from cropland during the time of year when leaching potential is high.

Phosphorus (P)

- ✰ It strengthens the root systems.
- ✰ It has a capacity for seed creation.
- ✰ It is disease resistance and pest prevention.
- ✰ It improves flowers and blooming.
- ✰ It strengthens tissues and flavors in edibles and veggies.
- ✰ Recommendations for phosphorus, potassium, and micronutrients depend on soil test values, yield goals, and crop rotation.
- ✰ If soil test levels are high, extra nutrients other than in-row starter fertilizer are not required for many crops/plants.
- ✰ Soil pH influences nutrient availability, particularly phosphorus.
- ✰ Soil pH should be adjusted to the level recommended for the crop/plant to be grown.

Potassium (K)

- ✰ It is important for roots and seed production.
- ✰ It supports the ability to tolerate extreme temperatures.

- It helps in extra prevention absorption of other minerals, like magnesium and calcium.

Calcium

- It strengthens and fortifies overall tissues everywhere.
- It helps in neutralizing acidity both in plants and in surrounding soil for optimal health.

Magnesium

- It helps in increasing the intake of phosphorus and boosts the production of chlorophyll.
- It gives a healthy green color and encourages the absorption of CO2.
- Plants that do have magnesium will have poor coloring and anemic appearance.

While plants need all of these nutrients for optimal health, higher numbers of some are more necessary at some times than others.

1.8.4 Deficiency in Plants due to Macronutrients

Nitrogen

- It is a mobile nutrient, so with its deficiency, plants move it from older foliage to younger with actively growing leaves.
- The older leaves become yellow first, while the new leaves remain green.

Phosphorus

- The entire seedling is stunted during early growth.
- As per the species, leaves may become dull green, yellow or purple-tinged.
- The purpling of leaves is a classic symptom, with no color differences in leaves.
- The purple color in this will not give a sign of new leaves.

Potassium

- Its symptoms appear in older leaves first.
- The older leaves start yellow at the edges with some green portion at the base.
- In later stages, leaf edges turn brown and may crinkle or curl and small necrotic spots appear.
- In such a condition, plants may wilt, in spite of ample water in the substrate.
- Due to severe deficiency, leaves will die.

Calcium

- Its symptoms include slow growth and dieback of bud or root tips.
- Seedlings will have stubby little roots with brownish discoloration, which is common in acidic soils.
- A well-developed root system with many fine root hairs, is important for calcium uptake.

Magnesium

- Such nutrients is deficient in coarse-structured and in acidic soils.
- In this, uptake gets blocked if there is too much potassium in the soil.
- Like nitrogen, magnesium is a mobile nutrient, so deficiency symptoms show initially in older leaves.
- In this, the plant leaves show very characteristic yellowing between veins or ribs and appear streaked.

Sulfur

- The deficiency will lead to plants stunted.
- As it is not a mobile nutrient, so symptoms will be displayed on younger leaves which are initially light green and eventually develops scorched and curled margins.
- With this, dry areas form along the margins which spread inward to leaf midrib.

1.8.5 Deficiency in Plants due to Micronutrients

Micronutrient deficiencies are difficult to diagnose, as more than one nutrient is missing. Only the most common symptoms are listed below:

Iron

- Deficiency is common on alkaline or calcareous soil.
- Younger leaves become yellow to white and dry up.

Manganese

- The tissue between the veins mottles, while the veins remain green and are surrounded by a band of green tissue.

Copper

- New leaves are yellow at the tips and often become twisted.

Boron

- The deficiency affects the terminal bud which yellows, dries out and dies.
- Plants grow slowly.

1.8.6 Application of Nutrient Sources

The proper timing of nutrient applications is important.

- ☆ Apply nutrient sources as close to planting or nutrient demand as possible, so that nutrients are taken up by plants quickly and not allowed to runoff into surface water or leach into groundwater.
- ☆ Avoid application of nutrient sources to frozen ground and during periods of high potential for leaching and runoff.
- ☆ Application in late fall or winter of any nitrogen source for a spring-planted crop should be avoided if possible.
- ☆ Avoid the application of nutrient sources to sensitive areas, wetlands, sinkholes, and steep slopes.
- ☆ Calibrate nutrient application equipment accurately to insure that recommended rates are applied.
- ☆ Accurate and uniform applications of nutrients are necessary to maximize the nutrient potential of the fertilizer materials.

1.8.7 Manure Applications Recommendations

Testing

- ☆ Manures vary tremendously in nutrient content as per animal species, rations and storage conditions.
- ☆ The nutrient content of manure can be determined by laboratory testing.
- ☆ Whenever possible, manure should be sampled at least 6 weeks before planned application to allow time for analysis and plan development.
- ☆ A consistent baseline for nutrient content needs to be laid down and based on analyses taken at least twice a year till the uniform value is confirmed, and then every second year thereafter to verify its consistency.
- ☆ If significant changes occur like feed, management, animals or storage, then new samples should be collected for nutrient analysis.

Application of Manure

- ☆ Nutrient applications should be made at times of year that will minimize N and P losses to water and N volatilization loss to the atmosphere.
- ☆ Crop utilization of nutrients in manure and bio solids is maximized, if these materials are applied in synchrony with periods of crop uptake.
- ☆ Storage of manure may be necessary to facilitate the appropriate timing of nutrient applications.
- ☆ Nitrogen-based applications of manure will cause phosphorus soil test levels to increase over time.

- Application recommendations for daily haul operations include consideration of the slope, crop and vegetative cover.

Storage Capacity

- Optimal utilization of nutrients in manure and other nutrient sources is difficult without the ability to store manure for part of the year.
- Improving storage capacity will minimize the potential for nutrient loss or runoff and improves the possibility of proper timing of manure applications.

1.9 IPM in the Nursery

1.9.1 Introduction

Pest Management in the nursery involves:

- Locating of any bugs.
- Determine the severity of problem-related to inspection of undersides of old and new leaves for check spreading of disease.
- Making a plan to monitor plants for a few days and to implement cultural tactics like composting.
- Consider what controls to choose like the use of organic pesticides after all other options.
- Keeping notes of when to spray and take note of pest problems.
- Considering increasing the diversity of the nursery covering flowers in the vegetable garden to attract beneficial insects.

1.9.2 Points to Consider in Integrated Pest Management

Floriculturist needs to keep in mind the following points regarding integrated pest management:

Prevention: Keeping a Healthy Nursery

- Plant varieties that are suited to the site.
- Know what insects and diseases are prevalent in the nursery area and plant resistant varieties.
- Making use of trap crops and row covers to exclude and divert nursery pests.
- Encourage beneficial insects.
- Interplant flowers to slow the spread of problems.
- Stake plants to keep them off the ground and dry.
- Water regularly, so plants aren't stressed by drought.

- ☆ Mulch to prevent splashing soil and pathogens onto plants.
- ☆ Rotating crops to prevent the problem from over-wintering.
- ☆ Remove and dispose of diseased or infested plants.
- ☆ Remove all plant debris in the fall, so there is no shelter for over-wintering garden pests and spores.

Monitoring: Know what you're Dealing with

- ☆ Monitor the nursery plants regularly by watching for first signs of garden pests, like holes, wilting, webs and color changes and checking the undersides of leaves.
- ☆ Identify the problem before treating it by knowing disease, insect, nutrient deficiency, rabbit damage, *etc.* and spraying insecticidal soap.
- ☆ Identifying Plant Nutrient Deficiencies
- ☆ Some problems are seasonal, so squash borers will mature and leave the garden by mid-summer.

Analysing: Amount of Damage to be Tolerated

- ☆ Is the insect making the leaves ugly and not harming the flowers?
- ☆ Knowing whether the problem is gone before any real damage.
- ☆ Floriculturists should understand, whether the problem is isolated which will clear, if the weather changes or the insect moves on.

Control: Stopping the Problem before it Damages

- ☆ When the problem won't go away on its own, floriculturist should start with the least toxic solution first.
- ☆ Trapping of garden pests with diversion crops, lures or sticky bands, is useful for lowering pest populations and for monitoring the problem.
- ☆ Hand removal is easiest, if it started early.
  - ❖ Removing infected plants before it has time to spread.
  - ❖ A gardener can monitor for pest egg masses on the undersides of leaves and squash or remove the eggs before they become a problem.
  - ❖ Many beetles are slow-moving, especially while mating, so the gardener needs to knock them off plants into a jar of soapy water.
  - ❖ Borers can often be cut out of the affected plant part without killing the plant.
- ☆ Pesticides use may become necessary. There are dozens of organic and botanical pesticides available for use. A gardener should start with the one that is least toxic and most effective as the product is organic and is not harmful.

1.9.3 Nursery Pests and their Controls

Aphid

- ✰ Apply strong spray of water or insecticidal soap.
- ✰ Allow beneficial insects like lady bugs or lace wings.

Flea Beetle

- ✰ Use agricultural row cover to protect young crops.
- ✰ Sticky traps help to monitor the presence of beetles.

Cucumber Beetle

- ✰ Cucumber beetles eat roots, leaves, and flowers and transmit bacterial wilt disease along the way.
- ✰ Control of damage done by cucumber beetles by buying wilt-resistant plants and keeping plants off the ground by using a trellis.

- Many flowers are sensitive to many chemicals, hence gardener should read the label's instructions carefully and use chemicals only as last resort.
- Use agricultural row cover when crops sprout and remove the cover when flowers appear for pollination.

Imported Cabbage Worm

- Spray Bacillus thurengiensis at first sign of cabbage worms.

Leaf Hopper

- Row covers
- sticky yellow traps
- insecticidal soap

Stink Bugs and Harlequin Beetles

- Handpick especially eggs.

1.9.4 Care of Nursery Plants

Insects and diseases are attracted to stress, damaged or otherwise unhealthy plants, so for a gardener, the main aspect is preventive control, which involves taking good care of plants by paying close attention to them and providing healthy, vigorous growth conditions. For giving healthy, vigorous growth conditions to garden plants, a florists' should:

Growing Plants in Healthy Soil

- Add organic matter to the garden every year to improve nutrient levels, soil structure, and water-holding capacity.
- Make sure that plants are getting water and nutrients as needed, supplemented with organic fertilizers if necessary.

- ☆ Don't force plants to compete for water and nutrients.
- ☆ Control weeds with mulches like landscape fabric or plastic and make a habit of pulling a few weeds every time on visiting the garden.
- ☆ Thin the seedlings so that they are not overcrowded and there is good air circulation between them.

Making Annual Garden Cleanup Part as a Daily Routine

- ☆ Leaving old squash vines, tomato plants and similar debris in the garden after the harvest ends is like putting out a mat for pests and pathogens.
- ☆ Many insects overwinter in such debris and get the early start nibbling on plants during the spring season.
- ☆ Many plant pathogens also live in the soil year-round.

Removing and Disposing of Diseased or Infested Plants

- ☆ The gardener should turn other debris into the soil or put it in the compost pile.
- ☆ Loosen the soil with a fork or spade so that any remaining eggs, larvae, or pupae will be exposed to birds and cold temperatures.
- ☆ Cut or remove the weeds around the garden as they can also harbor insect pests.

Rotating Garden Crops

- ☆ Many insects and disease-causing organisms overwinter, in the soil near their host plants.
- ☆ If you grow the same plant in the same place the next year, you will give those pests a big head start.
- ☆ Crop rotation will reduce insect damage and minimize exposure to soil-borne disease organisms.
- ☆ Wait at least two years before planting the same or related crops in the same spot.

Crop Rotation in Balancing Soil Nutrients

- ☆ The first-year planting of heavy feeders which actually return some nitrogen to the soil.
- ☆ During the third year, a gardener should allow the soil to rest by planting light feeders.

Encouraging Diversity

- If you place smaller groups of plants throughout the garden, instead of planting all plants in one place, it will be more difficult for pests to converge on a whole lot.
- Interplanting herbs and flowers is another effective way to protect the garden.
- Many plants like marigolds and nasturtiums repel insect pests while dill, mint, and fennel, attract beneficial insects that prey on garden pests.
- Mixing some of these plants in the flowers gardens will encourage a more varied and naturally healthy ecosystem.

1.10 Basics of the Flower Propagation

1.10.1 Introduction

Flowering plants exhibit many variations in shape, size, structure, mode of nutrition, life span habit, and habitat. In this, the shoot system is differentiated into the stem, leaves, and fruits. Here the flower is a modified shoot that can be applied for sexual reproduction. After fertilization, the ovary is converted into fruits while ovules converted into seeds. The floral characteristics form the basis of classification and identification of flowering plants which can be seen by description of families, so flowering plants are shown in definite sequence using scientific terms.

1.10.2 Propagation Media

A good propagation medium is made up of components that provide optimum aeration, drainage, and moisture-holding characteristics, which are made from combinations of peat moss, perlite, vermiculite, sand or similar materials. The main role of the propagation medium is to provide support and moisture during plant development. Such requirements are different from those of potting medium, having sustained mature or growing plants over a longer period of time.

It is noted that many plants get easily root in water, but the roots that form can be extremely fibrous and stringy. Plants rooted in water will have a difficult time establishing after get transplanted to the container.

- Moisture: The propagation medium should be thoroughly moistened before use. Many organic materials, like peat moss, have a waxy outer coating that resists wetting. Floriculturists should make sure that water should be applied slowly to obtain uniform distribution which uses 2-3 applications. It is not uncommon for a medium to look wet on the surface but to be powdery dry in the middle. A well-moistened media will make it easier to stick cuttings later on.

- Light: Light is an important environmental factor in flower propagation. Low light levels cause the flower to root slowly while high light intensities can lead to stress cuttings, which causes them to burn or drop leaves. Diffused sunlight generally provides enough light for optimum rooting without causing injury to the cuttings.
- Humidity: Since cuttings do not have roots, they cannot replace the water lost through transpiration, so floriculturist/gardener needs to maintain high humidity around the cuttings to cut down on the amount of moisture lost to the atmosphere. Such conditions can be provided by placing a clear piece of plastic over the propagation area to cause condensation so as to form on the underside of plastic which gives required humidity.
- Ventilation: Adequate ventilation is required to avoid disease problems, so the gardener should use plastic covering such that air can flow freely around the cuttings as they root.
- Temperature: For good results, the day temperatures should be around 21°C. In winters, soil can be as much as 2-5°C degrees less than the air temperature, so bottom heat should be given if possible. The ideal root zone temperatures for many flowers should be around 21-23.8°C.
- Rooting Hormones: Rooting hormones are often used to promote root formation. Such materials provide supplemental auxin, which is a naturally occurring plant hormone used for root development. In this, the basal end of the cutting is dipped in the chemical prior to sticking it in the propagation medium. Such products come in different strengths and will vary as per the type of flower being propagated.
- Stem and Section Cuttings: There are two types of stem cuttings in flowers which are tip cuttings that cover apex or plant tip with a small portion of the stem, while other is section cuttings that cover 2-3 inch section of stem.

1.11. Nursery Tools and Implements

1.11.1 Basic Tools

For doing day-to-day routine work in the nursery, the gardener uses different tools, implements, and accessories. Among these garden tools, some tools are simple and are used for simple operations while for doing specific operations, gardener/ nursery uses specific types of equipment such as:

Kudali

It is an important tool used for digging of soil, pits or basic digging work before preparation of nursery beds. It has a metal blade attached to wooden handle and is operated manually.

Garden Fork

It is used for breaking of soil clods and separation of medium size soil during preparation of nursery bed or fields.

Weeding Fork

It consists of long handle with a blade of handle teeth used for loosens the soil and helps in weeding. It is drawn manually using the handle to collect the weeds and cops of plants *etc.*

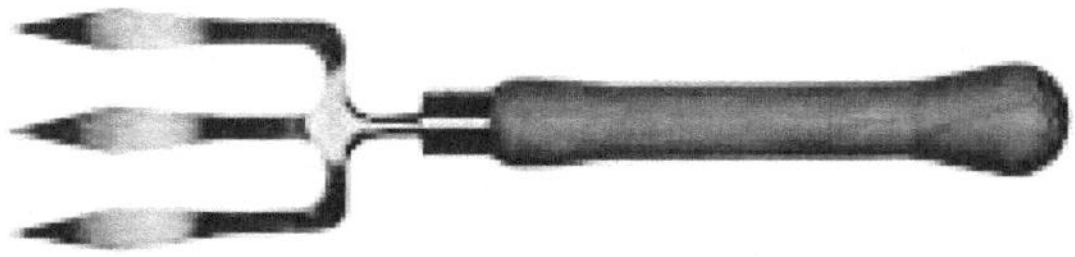

Crow Bar

It is an iron rod with one end pointed and other as a wedge shaped. It is used for digging out large or hand boulders from the soil and digging of pits. It is also used for breaking hard soil pan.

Shovel

It has an iron blade of spoon shape and wooden handle and is used in field transportation of dug soil for levelling of fields for preparation of nursery beds.

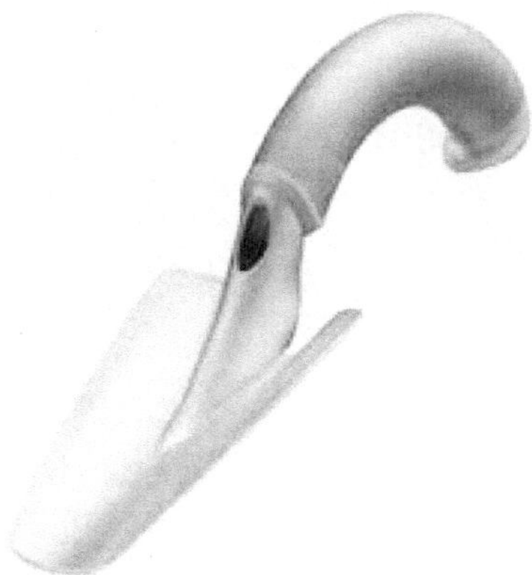

Pick-Axe

It is made of carbon steel having two edges with provision of axial hole for attachment with handle where one edge of pick axe is pointed while other edge is broadened. It is used for digging hard, compact and stony soils.

Spade

It is used for lifting and turning the soil and also for digging the pit, preparing channel for irrigation and drainage lines.

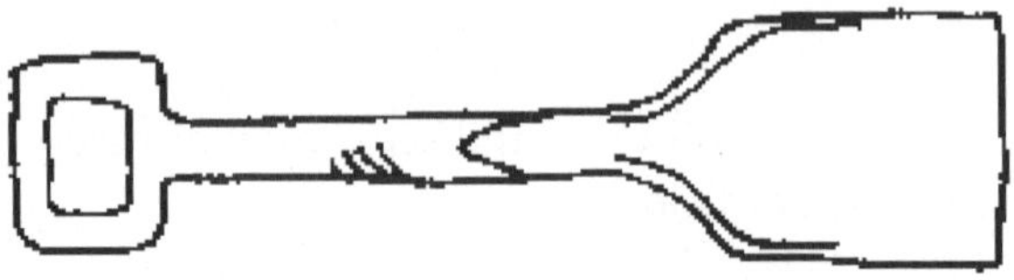

Hoe-cum-Rake

It is rectangular shaped metal blade with fork like fore edge used for digging, hoeing, earthing, levelling and collecting weeds.

Furrow Opener

It is used for opening narrow and shallow furrow after sowing seeds in nursery.

Hand Leveler

It consists of narrow rectangular metal blade attached to long wooden handle.

Trowel (Khurpi)

It is of shovel shape but small in size with iron blade and wooden handle. It can be made in many shapes as per local designs and requirements. It is used for hoeing, weeding and nursery plants and also for transplanting seedlings.

Axe

It has an iron blade which is fastened to wooden handle and is mainly used for felling trees and cutting branches.

Bill Hook (Darat)

It is made of iron curved at the far end which is used for cutting hardy branches of plant and other woody shrubs in the field.

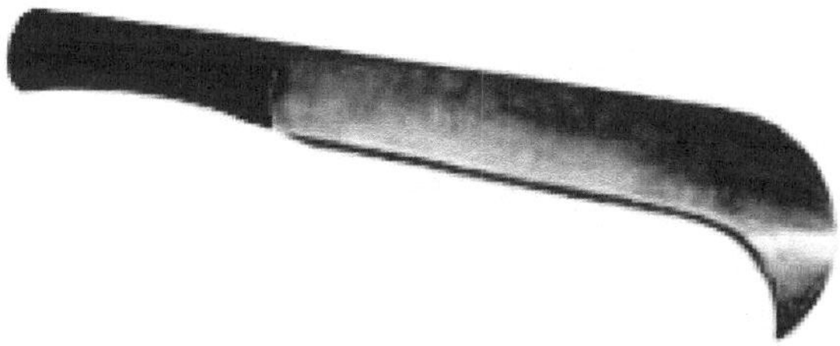

Sickle

It is used for cutting grass and leafy vegetables.

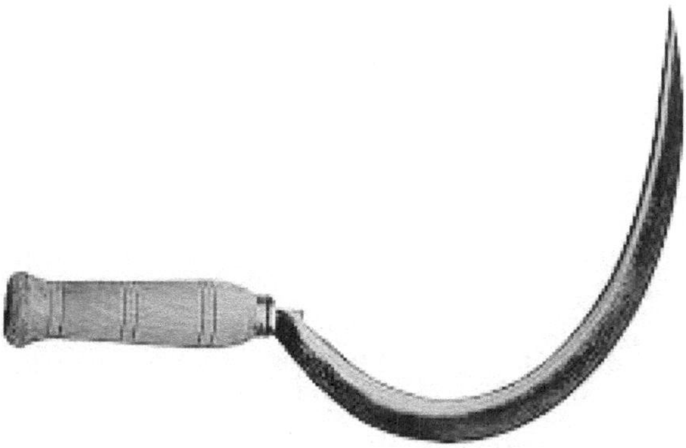

Wheel-Barrow

It is manually operated small trolley used for carrying nursery plants, compost, fertilizes, leaf litter, horticultural produce, stones *etc.* from one place to another place.

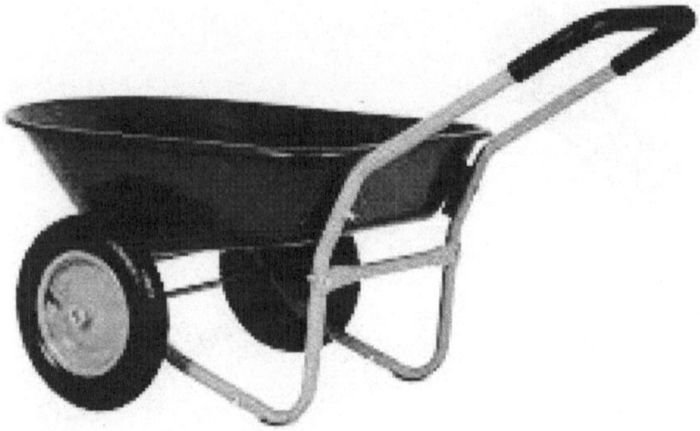

Cultivator

It can be tractor or hand driven implement used for tilling the soil efficiently. It has tynes of quality carbon comprising of different sizes, depending upon the plough depth.

Hand Driven Implement.

Disc Harrow

It is used for pulverizing the soil and for turning crop residues, weeds and other debris into the soil. Also it can be used for deep ploughing and turning of soil.

1.11.2 Grafting and Budding Tools

Knife

Knives has combined blade which can be used for grafting and budding having 7.5cm long blade and strong long handle. It has spatula at the end of handle, which is used for lifting the bark during budding operation. Sometimes knife has two parallel double blades, used specially for lifting or removing the patch of a bud from the budwood.

Secateurs

It is the most important tool for propagator or nurseryman which can be used for removing scions, lopping off rootstock, preparing scion sticks, removing undesirable shoots/sprouts from stock and training and pruning operations.

Grafting Machines

Many machines have been developed and commercially used for the preparation of scion sticks and bud wood for budding and grafting operation in many developed countries.

Pruning Saw

In nursery, many types of saws are required for doing different tasks involved in propagation of horticultural plants which can be crescent saws, tapered saws and straight saws. All these saw have long and wide set teeth to facilitate pruning or cutting of green wood.

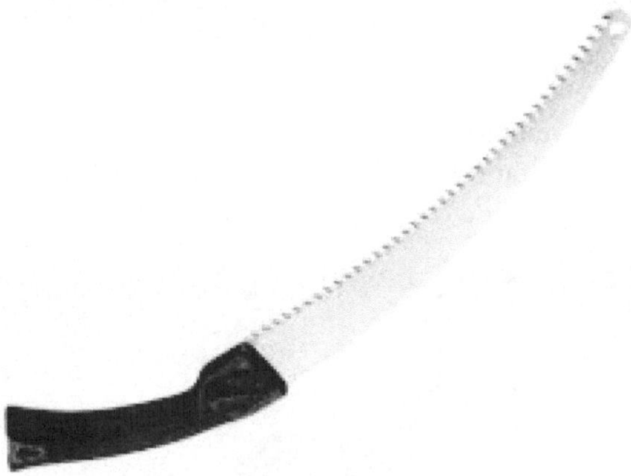

Grafting Wax

Wax is used by propagator to seal the graft union for preventing moisture loss and desiccation of cells, at cut surface and to prevent the decay of wood, by way of checking the entry of pathogens. Waxes can be hot wax or cold wax.

Pots

Pots of different shapes and sizes are used in nursery. They are made from clay, metal or plastic. Pots are of different types used for potting ornamental plants, fruit plants, saplings *etc.* Iron pots/plastic pots are used for irrigation purpose also.

Spray-Pumps

It is used for spraying insecticides/pesticides/fungicides, to remove infection of pathogens or insects. These are of different shapes, size and types.

Chain Saw

It operates by fuel (petrol/kerosene) which is used to cut wood logs of bigger size and lopping of branches and shoots.

Chain Weeder

It is operated by petrol/diesel/kerosene used to slash the weeds on large scale.

Rose Can

It is a tubular pipe with rose which is fitted on can by which water gets sprinkled over nursery beds, till germination of seeds and to avoid splashing of seeds from nursery beds, due to loose pipe irrigation and flood irrigation.

Iron Pan

It is made of iron and used for transporting pot mixtures, potted plants through head load from one place to another for short distance.

Hose Pipe

This is available in convenient length. Irrigation to nursery plants is made possible to any extent.

Pruning Shear

It is made of iron fitted with wooden handle which is used to prune unwanted branches, collection of scion, trimming of edges and hedges and many topiary work.

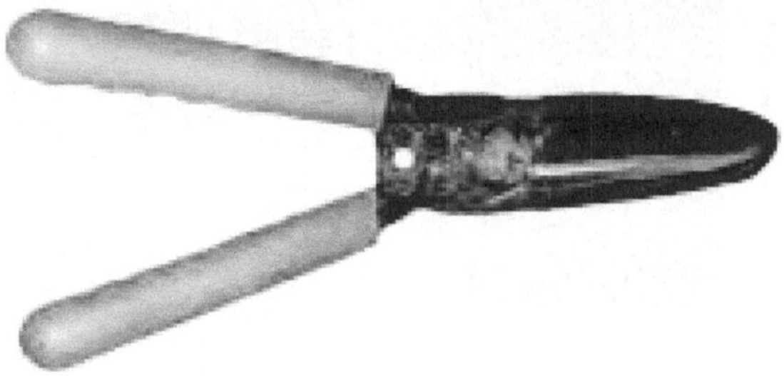

Scythe

It is long knifed, fitted with wooden handle which is used for slashing of weeds.

1.12 Use of Growth Regulators in Seed and Vegetative Propagation

1.12.1 Growth Regulators

- Plant Growth Regulators are small, simple, chemicals produced naturally by plants to regulate their growth and development.
- They are organic compounds, other than nutrients that modify plant physiological processes.
- They are also called biostimulants or bio inhibitors that act inside the plant cells to stimulate or inhibit specific enzymes or enzyme systems.
- They help in regulating plant metabolism.
- They normally are active at very low concentrations in plants.
- The effectiveness of growth regulators not only depends on concentration but also on the method of application.
- In this, auxins serve as an effective and widely used rooting hormone.
- Among synthetic auxins, IBA and NAA are found to be most effective for inducing rooting.
- The regulators can be of a diverse chemical composition like gases (ethylene), terpenes (gibberellic acid) or carotenoid derivates (abscisic acid).
- Plant Growth Promoters promote cell division, cell enlargement, flowering, fruiting and seed formation, which include auxins, gibberellins, and cytokinins.
- Plant Growth Inhibitors are chemicals that inhibit growth and promote dormancy and abscission, in plants which include abscisic acid.

1.12.2 Factors Affecting Plant Growth

There are certain factors that affect the growth of plants which includes:

Light

- Plants are autotrophs and require light for making their food.
- Limited light or absence of it greatly affects the growth of the plant.
- The intensity of light, quality of light, and light duration influence the movement of stomata, chlorophyll synthesis, photosynthesis, and various other physiological factors.
- Light also encourages flowering and fruiting.
- During winters when the days are short, the growth of the plants is retarded.

Water

- ☆ Plants cannot survive without water.
- ☆ Around 90 per cent of the plant, body comprises water.
- ☆ Plants become stressed in the absence of water and die.
- ☆ Water present in the soil is absorbed by the plant, which absorbs and transports the nutrients along with it.
- ☆ Water keeps the plant hydrated.

Temperature

- ☆ Plant growth is greatly influenced by temperatures.
- ☆ High temperatures speed up transpiration, photosynthesis and germination processes.
- ☆ Low temperatures slow down the growth of plants.

Nutrients

- ☆ Plants require proper nourishment for their growth and development.
- ☆ Soil nutrients can be macronutrients and micronutrients.
- ☆ Nitrogen, potassium, calcium, magnesium, sulfur, and phosphorus are macronutrients required by plants.
- ☆ The micronutrients include iron, copper, *etc.*
- ☆ The deficiency of nutrients in plants makes them prone to several diseases.
- ☆ Even if a single nutrient is lacking, it results in stunted growth of the plant.

1.12.3 Application of Growth Regulators

- ☆ Plant Growth Regulators is used to modify crop growth rate and growth patterns, during the various stages of development, from germination through harvest and postharvest preservation.
- ☆ In this, growth-regulating chemicals have positive influences on major agronomic crops, which can be of value.
- ☆ Applied plant growth regulators may influence the growth, development, and survival of woody plants.
- ☆ In this, plant response varies not only with growth regulator and dosages applied with method and time of application, but also with plant species and genotype, stage of plant development, site, and environmental conditions.
- ☆ Their results success in plant growth with some growth regulator treatments, which involves continued research, with an idea to extend the use of regulators in commercial practice.

- ☆ The Effect of Plant Growth Regulators on yield results in the prevention of lodging in cereals, prevention in preharvested fruit drop, synchronizing maturity to facilitate mechanical harvest, hastening maturity to decrease turnover time and reducing labor requirements.
- ☆ The growth regulators have been used by horticulturists to increase new growth-regulating compounds and its effect on the amount and quality of food and fiber crops of woody plants.
- ☆ In plants, growth substances are one of the effective means of delaying the senescence of leaves and retarding the abscission of reproductive organs.
- ☆ Certain growth-regulating chemicals like triacontanol, which influence photosynthesis, nutrient uptake, enzymatic activity, and gene regulation are considered beneficial to various crops.

1.12.4 Effects on Crop Growth

- ☆ Several plant hormones have been shown to affect the germination of seeds of some plant species.
- ☆ The primary event of breaking seed dormancy is stimulated by gibberellins.
- ☆ Volunteer plants of field crops, which begin growing after harvest, attest to the fact that PGRs that break dormancy is not required to aid germination.
- ☆ Germination of field crops decreased by cold soil temperatures.
- ☆ The cotton seeds soaked for 6 to 24 hours in gibberellic acid or cyclic AMP increases germination percentages over a range of temperatures.
- ☆ In this, seeds absorb the PGR in the embryo cells.
- ☆ For the absorption of PGRs to occur, seeds must be placed in a solution containing the PGR for eight or more hours.
- ☆ The methods of seed treatment cause little seed response as the PGR did not penetrate the seed coat.
- ☆ While germination of several crop species has been increased by gibberellic acid, indole acetic acid, succinic acid and fusicoccin, it needs seed soaking to incorporate the chemical into seed.
- ☆ Soaking initiates water imbibition, germination and softening of seed, which makes planting with current planters difficult.
- ☆ For most field crops, poor seedling emergence by soil crusts, dry soils, and cold soils is more of a problem than poor germination.
- ☆ Large seed broadleaf crops emerge by elongating hypocotyls dragging large cotyledons by the soil.

- ☆ Some short statured wheat varieties have short coleoptiles and cannot be planted at soil depths greater than 2 inches which prevent their use in dry soils, where seeds must be planted deeper than 2 inches to come in contact with moisture.

1.13 Selection and Maintenance of Mother Trees

1.13.1 Introduction

A mother plant is a plant that is grown specifically to take cuttings from to make clones. Mother plants are generally considered to be special plants with unique characteristics that gardeners want to reproduce. Cuttings taken from a mother plant will contain many or all of its mother's desired characteristics.

1.13.2 Growing Features

Mother plants are often grown from seeds and exhibit unique and desirable characteristics.

Cuttings taken from the mother plant will grow to have the same genetic qualities as the original plant. This is why the plants resulting from these cuttings are called "clones." When a stem cutting is taken from a mother plant, the piece of stem will produce new roots if it is properly hydrated. Additionally, some plants can be grown from leaf pieces, or leaf cuttings, and these will produce both stems and roots.

Some advantages of growing and maintaining a happy mother plant are the clones from it will have several positive aspects over plants that are grown from seeds.

The gardener should know the exact type and amount of crop the plant will produce. Another advantage is that, when the plant is cut and grown properly, growers/ gardeners can see a 100 per cent success rate in propagating new plants. Taking clones from a mother plant, is considered to be an economical way of producing new plants.

Cloning is best done in vegetative stage of growth for all plants. For this, a grower/ gardener needs to set up a light, which should be given for 18 hours for mother plant to keep her ready for vegetative propagation stage. Mother plants needs regular schedule for nutrients and water. Overfeeding will result in locking out required nutrients, which mother plant needs to thrive. Over watering can cause a multitude of mold, fungus, and bug problems. An over flux of food or water in your mom's system, when it comes time to take cuttings, can also make rooting more difficult. Giving mother plant right food before cloning, can make a world of difference in their babies. Gardener should see that whatever is in mother plant system at the time of cutting, will also be in the system of her clone.

1.13.3 Selecting Plant as per Mother Flowers

Gardener should understand how a plant will produce quality clones, for this, the gardener has to look at the plant itself, for having clones of hardy plant which matures as per schedule. Instead of choosing a clone as your mother plant, it is better for gardener to grow mother plants from a seed, which will help in controlling all aspect of environment from the beginning.

Similar to humans, every plant is genetically unique, so two seeds from the same plant can exhibit very different phenotypes or traits.

Selecting a plant with right traits to clone is basically a matter of personal preference, so to choose a mother plant, pick one that fits the preferences and which exhibits universally positive traits like:

- Strong roots
- A strong stem
- Resistance to pests and fungi
- A large harvest yield
- Steady, easy maturation

When the plant reaches the vegetative stage, in such case, the gardener has to take the clippings of all and should label them, so that he should know, which is which. Gardener won't be able to sample the harvests, till the plants reaches flowering stage, and by then they're reaching the end of their life cycle.

Once the gardener determined about the plants, which he wants to populate in garden, he needs to select its clipping to be as mother plant, which spends its entire life in vegetative stage.

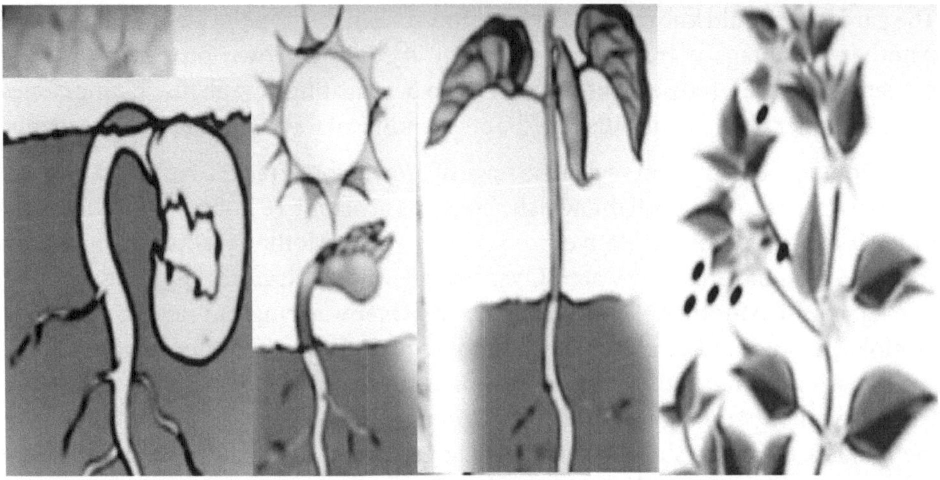

Life Cycle.

Mother plants can live much longer than plants that are permitted to progress through their natural life cycles. Some plants live for only one season, while others progress through their life cycle, year after year, with dormant periods between these cycles. A well-kept mother plant can potentially live for years, continually producing new branches that can be cut and grown into clones.

1.14 Collection of Scion Wood Stick

1.14.1 Scion Wood Stick

Scion is a piece of vegetative material taken from a tree that produces fruit variety which a gardener/florist wants to graft. For grafting, a gardener has to decide what variety he/she requires with taste, season, qualities and suitability for particular site in terms of pest and disease resistances and vulnerabilities.

Shoot or twig, also known as scion or scion wood is normally collected in January or February and can be stored under refrigeration, till the grafting season. Grafting is the process by which a fruit tree is asexually propagated, which results in new young tree that can be genetically identical to the parent tree.

1.14.2 Selection Factors

There are certain factors which should be considered, while selecting a scion wood such as:

Time Period

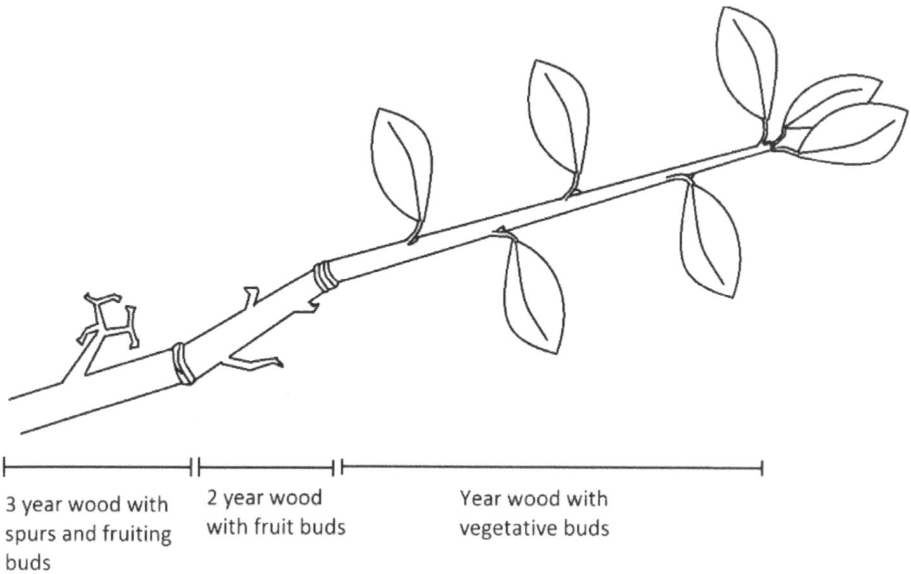

☆ When selecting the scion wood, gardener/florist should find the one year wood as the year-wood will not normally have any fruit buds, which are usually found on the wood, that is two years or older.

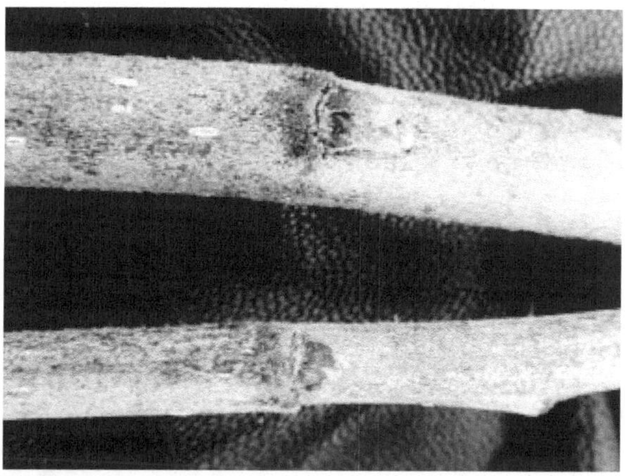

☆ To the graft using material from old tree, a gardener needs to find such which doesn't have huge amount of new growth, from where he/she cuts the scion material.

Timing

- Scion wood should be cut in the late winters or early spring when tree is not in active growth and before bud-burst.
- Flowers trees are dormant through the winter months, so material can be collected any time until about February.
- The best time to cut scion is just before gardener graft with it as scion wood can lose its vitality as it dries out over time.
- If the scion material is cut early in winter, then gardener needs to store it longer in cold storage, sealed in plastic bag.
- Due to this reason it is advised to cut scion material in late winter rather than early, to minimize time in storage.

Size and Shape

- The size, shape and number of the scion wood twigs to be collected will depend on the type of usage and work.

- For the bench grafting, collect straight growth which should be of pencil thickness.
- If rootstock and scion material are of same thickness, it will allow more cambial contact which increases the likelihood and strength of graft join.
- Cut these right back to just before the growth scar so that the end gets dried out and be cut off before grafting giving much of year-wood to work.
- For the bench grafting, if 3-4 buds are used, it should depend on length of new growth on tree which new trees to be grafted from each scion.
- For framework grafting, more scions are required as per the size of tree where material scions of 7-9 buds long are used.
- When cutting the wood for frame working, don't avoid bent or curved new growth, as these can actually be useful to graft new branches to framework in direction as required to grow.

Process

- Good scion wood should always be one year old wood or the wood that grew in the previous summer.
- Such wood can usually be found on the outermost ends of limbs or water sprouts from the trunk or large branches.
- To avoid water sprouts which comes from near the ground and below the old graft union, such sprouts can be from unknown root stock.
- Cut the new growth into pieces which should be cut about pencil/pen length and should have a diameter of a pencil or ¼ inch.

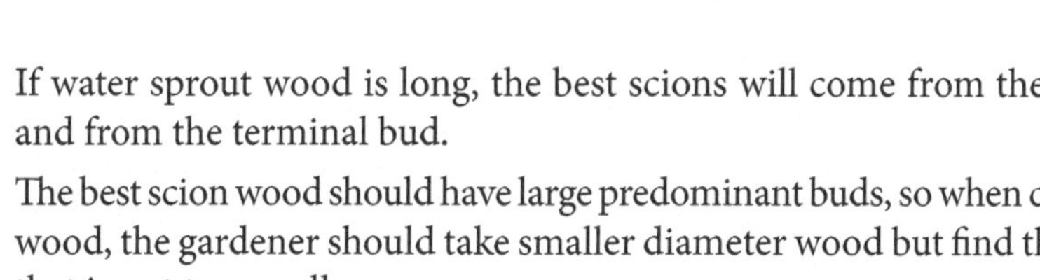

- If water sprout wood is long, the best scions will come from the middle and from the terminal bud.
- The best scion wood should have large predominant buds, so when choosing wood, the gardener should take smaller diameter wood but find the wood that is not too small.
- The process of grafting is quite simple where the selected piece of scion wood is inserted in the rootstock of a young apple tree where it will heal and fuse together to produce a new tree.

- ☆ The most important point in collecting scion wood is to be sure to collect new growth, twigs or shoots that emerged the previous summer.
- ☆ New growth is identified by its smooth, reddish or greenish bark in contrast to older growth which will have rough, grayish-colored bark.
- ☆ New growth will have small, tight buds and rarely have side limbs, twigs or branches which are common on older growth.
- ☆ On older, unmanaged trees, new growth may be difficult to find, so such an approach can be found out on tips of twigs high up on the sunny side of the tree.
- ☆ New growth can be found out on water sprouts which are vigorous, whip-like shoots ranging in size from a couple of inches to several feet in length and typically found growing vertically upwards from the trunk or larger limbs.
- ☆ Not all water sprouts are new growth, so a gardener has to make sure that any water sprout selected for scion wood has no side branches or twigs.
- ☆ Collect proper scion wood cuttings in late January to mid-February which should be ideally 8-10 inches in length and of approximate diameter of pencil/pen.

Label and Storage

- ☆ After cutting the scions, it is important to bundle them up and label them before storing them.
- ☆ Having cut your scion material whilst dormant, you need to keep it this way until you graft it.
- ☆ To store your scion wood, you have two main options, refrigeration or 'heeling in' until they are needed.
- ☆ If storing in the fridge, wrap the scions in damp paper towels and seal in a plastic bag to lock the moisture in.
- ☆ Make sure there is no fruit in the fridge as this will reduce the success rates of stored scions as fruits naturally release ethylene, a hormone that promotes ripening.
- ☆ If you don't have refrigeration capacity, then heel the scions in ditch or large container filled with damp sand and soil mixture and place them against the north-facing wall.
- ☆ If the method of scion storage is selected, then cut the scions longer than usual, as the base will slightly dry out and become useless for grafting.

1.15 Techniques of Propagation through Specialized Organs, Corm, Runners, Suckers

Vegetative plant structures that arise from plant stems include rhizomes, runners, bulbs, tubers, corms, and buds.

1.15.1 Propagation by Specialized Organs

- ☆ Plants reproduce to perpetuate their offspring by sexual and asexual means of reproduction.
- ☆ The Sexual reproduction method produces offspring by fusion of gametes, resulting in offspring genetically different from parent plants due to genetic exchange occurs during fertilization which came from both parents.
- ☆ Asexual means of reproduction produces new individuals without the fusion of gametes, which is genetically identical to parent plants, except when sudden change, a mutation occurs.
- ☆ The plant is composed of roots, stems, leaves, and flowers, which can be used as material for propagation. The commonly utilized form of plant reproduction by people is seeds.
- ☆ The asexual methods include cuttings, grafting, budding, layering, division, separation and micro propagations that uses when seed propagation is not feasible.
- ☆ Asexual methods of plant propagation multiplies cultivars with individual desirable characteristics that do not come to true from seeds, to ensures faster initial plant growth and higher survival rate of plant, to produce higher yield with quality product, to reproduce plant which has a short life span and seed dormancy and to propagate a plant which is sterile to reproduce by seeds.
- ☆ In many plant organs, modifications exist to enable natural vegetative propagations where the stem is the most important one which produces buds that completely grows to a new plant with roots, stems, and leaves.
- ☆ All daughter plants produce from such organs are identical called clones of mother plant which serves as food stores that allow quick burst growth of plants in spring.
- ☆ The specialized vegetative structures include runners, suckers, crown, offsets, bulbs, corms, tubers, tuberous roots, rhizomes, and pseudobulbs are used primarily for the storage of foods, nutrients, and water during adverse environmental conditions and for propagules.
- ☆ Plants possessing these modified plant parts are generally herbaceous perennials, in which the shoots die down at the end of a growing season but the fleshy vegetative structure usually does not die and remains in the soil, which put forth new vegetative growth in the next season.

- ☆ Plants that survive as underground storage organs are called genotypes that can withstand a period of adverse growing conditions in their regular growth cycles and used as propagules.

1.15.2 Runners

- ☆ Runners are also known as stolons which are similar to rhizomes which exhibit horizontal growth at or just below the soil surface.
- ☆ Unlike rhizomes, stolons originate from existing stems.
- ☆ As runners grow, they develop roots and shoot from buds located at nodes or at runner tips.
- ☆ Intervals between nodes are more widely spaced in runners than in rhizomes.
- ☆ In this, new plants arise at nodes where roots and shoots develop.

To Propagate Plants with Runners

- ☆ Simply cut the new plant from the mother plant before the new root system.
- ☆ For taking runners, it is helpful to continue to root them in water for a couple of weeks.
- ☆ Transplant the new plants and keep the soil moist for 2-3 weeks.

Propagation by Runner

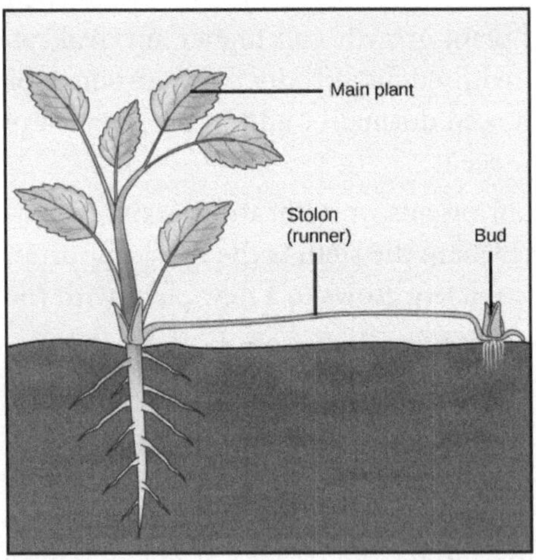

One plant may have several runners and one runner may grow several nodes.

The long and flexible runners are easy to bend and positioned according to the desire of the person cloning them. When runner separated from the mother plant with intact roots, it serves as units of propagation. Growing plants from a runner is the easiest and quickest way of propagating method. Runner is strictly a horticultural term; botanically they are simply classified as solons.

Corms

Corms are enlarged bulb-like underground stems. These vegetative structures store nutrients in fleshy, solid stem tissue and are typically surrounded externally by papery scale-like leaves. Due to their external appearance, corms are commonly confused with bulbs. The major difference is that corms consist internally of solid tissue, while bulbs consist of layers of scale-like leaves. Corms produce adventitious roots and possess buds that develop into new plant shoots. Plants that develop from corms include crocus and gladiolus.

1.15.3 Suckers

- Suckers or root sprouts are plant shoots that arise from buds on underground roots or stems.
- Suckers may also sprout from buds near the base of parent plant and can grow into new plants.

- A number of shrubs and trees propagate through sucker production such as roses etc.
- Plants can also propagate asexually by suckering.
- A sucker is a shoot that arises from adventitious bud on an underground root.

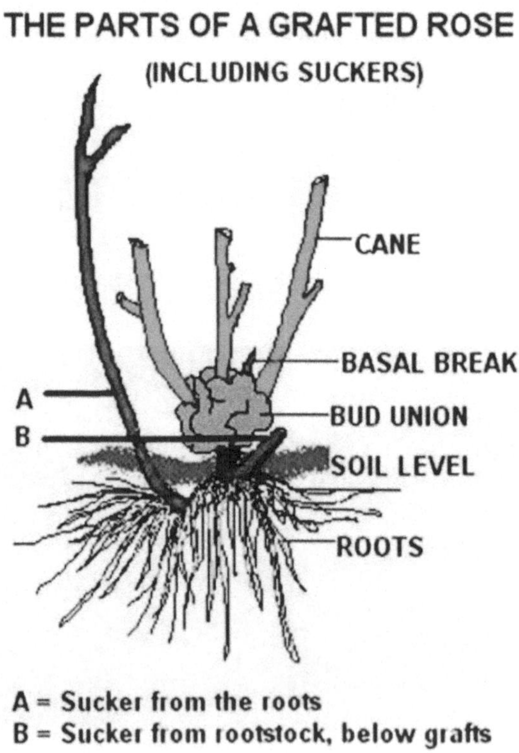

- As they have both shoots and roots, suckers can be used to propagate plants.
- To have more lilac plants, look for suckers that are at least two years old.
- Using a shovel, sever the root connecting the sucker to the main plant.
- Wait several months for the shoot to develop more feeder roots before transplanting.
- Suckers often arise from rootstock tissue on fruit trees and roses.
- Conscientious orchardists and rose growers know to prune off any suckers arising from the rootstocks, to keep the suckers from competing with the main plant.

1.16 Plant Protection

1.16.1 Introduction

Plant protection is the science and practice of managing pests, diseases, and weeds that damage crops and other plants, and which can have a devastating effect on farmer livelihoods. It focuses on keeping plants healthy from diagnosing diseases to implementing environmentally-friendly pest-management practices. With an ever-expanding population and increasing pressure on food and fiber supplies, protecting plants plays an important role in improving quality of life.

1.16.2 Process

The basic of plant protection takes place by anticipating emergences and spread of noxious organisms and in preventing them before they spread and result in agricultural pests for specific crops and regions, so for his, gardener/flower cultivator should analyze the following points:

Diagnosing Problems

- Ensuring to correct, diagnose of the problem, which can be done by proper identification of organisms that is responsible for the damage.
- Gardener/flower cultivator should study the taxonomy of invertebrates, micro-organisms, and plants, for knowing about their characteristics and behavior.
- Gardener/flower cultivators should collect and compare specimens for different resources.
- The Gardener/flower cultivator should use modern information technology and biotechnological tools to work in a lab for faster and correct identification of organisms.

Assessing the Extent of the Damage

- ✰ After collecting the specimen and getting information about the organism, the gardener/flower cultivator should assess the extent of the damage and yield loss which could take place from the organisms, so that the grower can take decisions about whether to invest resources in combating the pest.

Options Used to Control

- ✰ If the gardener/flower cultivator decides to take action to reduce pest attack, then he/she should know about different options available for controlling host plant resistance, cropping system and cultivation practices, conservation and introduction of antagonistic organisms like parasites and predators of pests by using chemical or biological-based pesticides.

Steps to be Taken

For protecting plants, gardener/flower cultivator should:

- ✰ Use bio-pesticides instead of chemical pesticides.
- ✰ Assess the Pest Defender ratio before using any pesticide.
- ✰ Use Agro Eco System Analysis based Integrated Pest Management.
- ✰ Grow crops that surround the main crop which attracts friendly insects, so that harmful insects can be killed.
- ✰ Resort deep plowing during the summer season.
- ✰ Use a light trap/sticky trap/pheromone trap for monitoring and mass trapping of pests.
- ✰ Use parasitoids and predators for biological control of insects, pests, and antagonists for diseases.
- ✰ Use chemical pesticides only if the above measures are not effective after taking advice from experts.

Precautions to be taken while Using Pesticides

If the Gardener/flower cultivator uses pesticides to protect plants from the pest, then he/she:

- ✰ Should follow all the prescribed safety precautions while using pesticides.
- ✰ Should wear safety gadgets like masks, hand gloves, *etc.* while spraying the pesticides.
- ✰ Should always spray in direction of the wind and keep themselves safe from the spray drift.

- Should always keep pesticides, plant protection equipment, *etc.* under lock and key away from children and pets.
- While purchasing pesticides, must see the packing and validity date of the chemical.
- In case of any pesticide poisoning, should immediately contact a doctor and should carry the empty container and information booklet of pesticide used.
- Should use pesticides as per the instructions on the label.
- Should dispose of used containers as per the instructions given on the leaflet.

1.17 Hi-Tech Nursery

Hi-tech nurseries use special techniques and methods to meet the demand for certain commercial plants. In this, plants are grown in a greenhouse, the building of glass or plastic tunnel, that are designed to protect young plants from harsh weather, by allowing access to light and ventilation.

Modern greenhouses allow automated control of temperature, ventilation, light, watering and feeding while some have fold-back roofs to hardening-off the plants without the need for manual transfer of plants to outdoor beds.

Such type of nurseries makes available to farmers high-quality seedlings. The seedlings instead of seeds are extensively used by farmers, which saves time and ensures no loss due to seeds not sprouting. The scientifically produced hi-tech seedlings will be healthy, generating high yields in fertile soil with quality irrigation facilities.

1.18. High Yield Varieties of Flowers

Flower production varies with cultivar or variety which depends on bulb size at planting time, the density of planting, cultural practices and climatic condition that prevails in the area. Flowers are ready for harvest in about 3 to 3.5 months after planting. Following are the flower crops variety that shows high yield:

1.18.1 Chrysanthemum

- Chrysanthemum which occupies a prominent place in ornamental horticulture is one of the commercially exploited flower crops.
- It is an important commercial flower crops of Karnataka having low yield and quality levels.
- It requires optimum dose of nutrients to standardize nutrient management for improving soil structure, physico-chemical properties and flower yield.
- In this, higher yield results from more number of flowers and heavy flowers.

- ✰ As flowering season varies from region to region, but average blooming season can be from July to February.
- ✰ The Chrysanthemum flower can be harvested 15 times, so good yield can be obtained from 10-11 tones of loose flowers per acre.

1.18.2 Gerbera

- ✰ It is commonly known as Transvaal daisy, Bar berton daisy or African daisy which is an important commercial cut flower crop.

- ✰ It is a colonial selection from mixed open pollinated seeds that is collected from germplasm of gerberas having dwarf, herbaceous perennial growth.
- ✰ It blooms throughout the year with peak flowering from May to June.
- ✰ An important cultivars of Gerbera includes Pre Intenzz, Stanza, Winter Queen, Cacharelle, Jaffa, Sangria, Diana, Thalsa, Sonsara, Paganini, Anneke, Nette, Rosetta, Gloria, Ginna, Ingrid, Pricilla, Alexias, Intense, Sunway, Zingaro, Balance, rosaline, dune and Monique.
- ✰ For commercial cultivation, such plants get propagated by sucker division and tissue culture systems.

- Such variety of flowers yields about 80 flowers per clump in an year which is good for cultivation in hilly regions.

1.18.3 Gladiolus

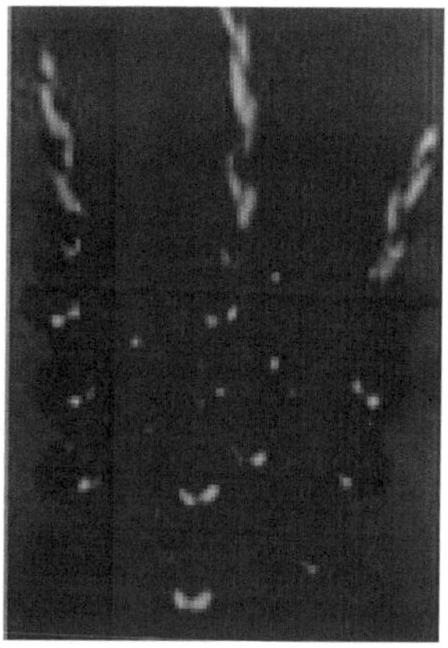

- Gladiolus flowers are liked by the people all over the world. It is a perennial flowering plant which has sword shape leaves, funnel shaped perianth and spoon shaped branches.
- The flowers bloom mainly in the month of October-March.
- It produces wide range of flower color ranging from pink to reddish, light purple to white, or white to cream or orange to red.
- It is also used to cure various ailments or diseases such as common cold, diarrhea, fungal infections, and meningitis and to get relief from constipation.
- For gladiolus plantation, land should be well ploughed before sowing.
- To bring the soil to fine tilth, ploughing must be done.
- Commercial propagation is through corms.

Gladiolus Corm

- Corms planted during September-October are ready for lifting during March-April in the North-Indian plains and up to August in hills.
- Corms are matured when 25 per cent cormels have become brown which generally take 30 to 45 days from flowering when the leaves also start yellowing.
- The yield of flower spikes and corms in gladiolus depends on variety, corm size, planting density and management practices.
- Gladiolus planted at a spacing of 30 x 20 cm yields approximately 1,50,000 marketable spikes per hectare.
- Additional income can be obtained from the sale of about 3.5 lakh corms.

1.19 Drought Tolerance

1.19.1 Drought Tolerant Plant

- A drought tolerant plant is that which can survive with little or no water other than available rainfall.
- Different locations and soil types affect the drought tolerance and different plants to vary with ability to withstand drought.

- Such plants are able to cope up with scarcity of water in various ways.
- Plants including trees, survive dry conditions with aid of large, deep root systems that tap water stored deep in the soil.
- As the plant roots are down deep and wide, they needs less or no artificial irrigation.
- Many plants developed leaves which resist drought conditions.
- Certain plants developed thick, leathery leaves that reduce water loss.
- Many plants have succulent leaves that store water and are waxy to prevent water loss.
- Some plants have hairy or fuzzy leaves where fine hairs keep moisture trapped at leaf surface to reduce evaporation.

1.19.2 Features

Drought resistant plants and flowers have:

- low water requirements
- easy-going soil requirements
- good sepals with worst soil needs

1.19.3 Types

Drought-Tolerant Annual Seeds

Many flowering annuals withstand long periods without rain which includes:

- African Daisy – Colorful flowers in shades of orange, apricot, yellow and white.
- Alyssum – Masses of tiny white, pink or purple flowers. Easy to grow from seed.
- California Bluebell – Native wildflower with cobalt-blue flowers from seed.
- California Poppy – Drought-tolerant flowers in a wide range of colors.
- Chinese Forget-Me-Not – Small, indigo-blue flowers. Makes a nice filler.
- Coreopsis – Native wildflower with yellow, gold or mahogany-red flowers.
- Cosmos – Late summer flowers in a variety of colors, heights and styles.
- Gaillardia pulchella – Scarlet-red flowers on a compact, bushy annual.
- German Chamomile – Small, daisy-like flowers make a great filler. Tea herb.
- Gilia tricolor – Native wildflower with delicate lilac and white flowers. Fragrant.
- Marigold – Popular annual with yellow, gold, orange and red flowers from seed.

- ☆ Mirabilis jalapa – Fragrant, trumpet-shaped flowers that attract hummingbirds.
- ☆ Silene aremeria – Vivid pink clusters of flowers. Drought-tolerant wildflower.
- ☆ Tithonia rotundifolia – Fast-growing annual with orange-red, daisy-like flowers.
- ☆ Zinnia – Brightly colored, drought-tolerant flowers for the late summer garden.

Drought-Tolerant Perennial Seeds

Many perennials show high tolerance for drought and perennials return year after year and spread out in clumps or clusters. They are ideal for hot, sunny slopes which get little water which includes:

- ☆ Aethionema schistosum – Rock garden perennial. Fragrant, pale pink flowers.
- ☆ Agastache (Hyssop) – Fragrant foliage and colorful flowers for hummingbirds.
- ☆ Alyssum montanum – Yellow-flowered, compact perennial for the rock garden.
- ☆ Aquilegia (Columbine) – Early spring-blooming flowers in a varicty of colors.
- ☆ Asclepias tuberosa – Native perennial for butterflies and hummingbirds.
- ☆ Aster – Spring or fall-blooming perennials with flowers in various colors.
- ☆ Baptisia australis – Native perennial. Bushy plant with blue, pea-like flowers.
- ☆ Callirhoe involucrata – Native perennial with reddish-purple flowers.
- ☆ Campanula rotundifolia – Short, spreading plant with violet-blue bellflowers.
- ☆ Catananche caerulea (Cupid's Dart) – Lilac-blue flowers on tall, wiry stems.
- ☆ Centranthus ruber – First-year-flowering perennial with reddish-pink flowers.
- ☆ Dalea purpurea – Native wildflower. Magenta-purple flowers for butterflies.
- ☆ Dianthus – Drought-tolerant perennials and biennials with fragrant flowers.
- ☆ Digitalis obscura – Sunset-red and gold flowers. Drought-tolerant foxglove.
- ☆ Echinacea – Long-blooming perennial with mauve or white flowers.
- ☆ Eryngium giganteum – Blooms in late summer with showy, silver-gray bracts.
- ☆ Gaillardia – Wine-red or scarlet-red flowers edged in yellow. Native wildflower.

- ☆ Gaura lindheimeri – Native perennial with pink-tinged, white flowers.
- ☆ Hollyhock 'Indian Spring' – Tall biennial with pink, white or rose-red flowers.
- ☆ Gypsophila pacifica – Perennial baby's breath with pale pink flowers. Filler.
- ☆ Iberis gibraltarica – Evergreen perennial with lavender-pink flowers. Rare.
- ☆ Lavandula angustifolia – Shrubby perennial with fragrant flowers and foliage.
- ☆ Liatris – Tall spikes of fluffy, purple flowers in late summer to fall. Native.
- ☆ Linum perenne – Drought-tolerant, native wildflower with sky-blue flowers.
- ☆ Oenothera (Evening Primrose) – Fragrant, yellow to orange-red flowers.
- ☆ Penstemon (Beardtongue) – Drought-tolerant perennial with tubular flowers.
- ☆ Perovskia atriplicifolia – Large, bushy perennial with lavender-blue flowers.
- ☆ Platycodon grandiflorus – Long-lived, drought-tolerant perennial. Lovely.
- ☆ Ratibida – Native wildflower with drooping, red or yellow or petals.
- ☆ Rudbeckia subtomentosa – Native perennial. Yellow daisies with brown centers.
- ☆ Rudbeckia triloba – Native perennial. Small, yellow daisies with black centers.
- ☆ Ruellia humilis – Native perennial. Lavender-blue, trumpet-shaped flowers.
- ☆ Ruta graveolens (Rue) – Blue-green foliage and greenish-yellow flowers.
- ☆ Salvia – Perennials with tubular flowers for hummingbirds and butterflies.
- ☆ Saponaria ocymoides – Bright pink flowers. For rockeries or groundcover.
- ☆ Scutellaria resinosa – Deep purple-blue flowers. Rock garden perennial.
- ☆ Shasta Daisy – Single or double, pure white flowers with yellow centers.
- ☆ Solidago nemoralis – Drought-tolerant perennial with lemon-yellow flowers.
- ☆ Thymus serpyllum – Drought-tolerant groundcover for pathways and rockeries.
- ☆ Verbascum phoeniceum – Tall spikes of flowers in pink, purple and white.
- ☆ Verbena bonariensis – First-year flowering perennial with rosy-purple flowers.

1.20. Transplanting

1.20.1 Introduction

- ☆ Transplanting is a method of planting seeds in seed box which on bigger seedlings gets transplanted from seed box to garden plot.

- It is done when seedlings attains 12 -15 cm height with development of 3 – 4 leaves.
- Transplanting is planting with the use of pre-grown seedlings or plants that had been propagated from seeds.
- Crop area is the space which is particularly used for growing crops.
- It is indirect as seeds are not immediately sown, but are first used to raise the seedlings in pots or in temporary place which are further used when reaches at right age.
- Field planting is done either manually or by mechanical means.

Direct seeding or transplanting depends on:

- Crop species to be grown
- Ease in planting and survival rate
- Gardener's familiarity
- Timeliness
- Financial capability of gardener/owner
- Return on investment

1.20.2 Preparing Land for Transplanting

When establishing new plantation, certain actions need to be implemented to ensure the long term success of plantation which involves initially land preparation that should be done before transplanting of plant material. The idea of land preparation is to provide necessary soil conditions which enhance the successful establishment of young offshoots or tissue culture plants that receives from nursery.

Factors

Factors to be considered for transplanting includes:

- Availability and quality of irrigation water;
- Field selection;

- ☆ Mechanical actions to be implemented;
- ☆ Chemical needs for pre-plant soil improvement;
- ☆ Tools and equipment needed for date cultivation;
- ☆ Labour needs;
- ☆ Irrigation design and installation;
- ☆ Leaching schedule;
- ☆ Hole preparation;
- ☆ Financial requirements
- ☆ Time schedule

Tools

Land and field preparation can be done by using:

- ☆ Deep working implements like ploughs
- ☆ Surface working implements like rotavator, harrows, hand hoes, khurpi, *etc.*
- ☆ Compacting implements like rollers, levelers, beams, *etc.*

Field Selection

The area selected for the establishment of the plantation can influence the cost of land preparation to an extent that it may not be viable to proceed with development, so gardener/florists should select the land for the establishment of new date plantation.

1. Availability of Water

- ☆ Although not always realized, any plantation needs more water for sustainable growth, so critical factors in terms of water for irrigation purposes includes:
    - ❖ Sustainability of water source
    - ❖ Quantity of water available for irrigation
    - ❖ Distance to the field
    - ❖ Quality of water

2. *Soil Depth*

☆ The plantation needs sufficient space for proper root development to support, so soil depth also influences the drainage and leaching possibilities and any obstructive layers should be evaluated in determining the root development.

3. *Soil Quality*

☆ As the plantation can grow and produce in different types of soil in both hot arid and semi-arid regions, where Adaptation could go from very sandy to heavy clay soil, hence soil quality is related to drainage capacity when soils are salty or irrigation water is characterized with high salt content. The optimum soil conditions are found where water can penetrate to at least 2 m deep. For evaluating soil quality, attention must be given to:

- ❖ Soil texture which influences water retention capacity
- ❖ Nutrient content to determine corrective measures necessary for soil improvement

4. *Soil Salinity or Acidity*

☆ Plant growth is influenced by either saline or acid soil conditions which, in the end, will result in loss of potential yield. Saline soils have electric conductivity of their saturated extract higher than 4 mmhos/cm at 25°C, with sodium absorption rate less than 15 and pH less than 8.5. Saline soils can be recognized by the presence of a white layer on the surface of the soil which results from the high salt concentration that can harm the growth and development.

Land Preparation Operation

☆ Once a suitable area for the plantation is selected, then planning operation gets finalized and actual preparation can be started.

☆ Such activities are divided into the structure and to speed up the implementation process for planting at the most suitable time, as per specific regional climatic conditions.

☆ Land preparation involves:

1. *Mechanical Field Preparation*

☆ The mechanical/initial soil preparation relates to the preparation of the field for further detailed preparation which includes irrigation system installation, whole preparation, *etc.* which involves:

- ❖ Debushing/bush clearing;
- ❖ Removal of stones and rocks;

- Ripping;
- Leveling of soil

2. ***Irrigation System Installation***
   - The type of irrigation system to be used will be determined by the availability of water, topographical and soil conditions. When initial soil preparation is completed, the installation of the required irrigation system will be implemented as per the prescribed design.
3. ***Soil Improvement***
   - The scheduling of soil improvement program depends on the grower, as certain applications could be combined with initial actions of soil preparation.
   - Due to the long waiting period, planting to the first production, it is a trend to establish plantations on new soils, with the exception of areas where the plantation is used for intercropping.
4. ***Hole Preparation***
   - The actual digging of the hole is done before the planting takes place, but it is seen that this is not the final preparation for planting operation itself as it is the point where required inputs like gypsum and organic materials are worked in soil.
   - For this, a hole of 1 m^3 be prepared and that the soil from the hole is mixed with the organic material and gypsum and soil mix is put back in the hole where the site is marked for positioning of small plants.

1.20.3 Procedure of Transplanting Seeds

To transplant seedlings, a gardener has to follow the steps shown:

- Use a hoe, spade or trowel to make a small hole in the garden area for each seedling.
- The hole should be deep enough so that the transplant is at the same depth in the garden as it was in the pot.
- Make the hole twice as wide as the root ball.
- Unpot a seedling by turning its pot upside down and cupping the seedling by hand.
- Gardener should keep the root mass and soil intact and if seedling doesn't come out easily, then he should gently tap on the edge of the pot or gently press on the bottom of each cell of the flat with fingers.
- Check the root ball's condition, if roots are wound around the outside of the pot, then work them loose with fingers so that they can grow out in the soil.

- ☆ Unwind larger roots and break smaller ones so that they all will point outward.
- ☆ Keep as much of original soil intact as possible and mix diluted liquid fertilizer in the soil of planting the hole so that plants should germinate faster.
- ☆ Keep each prepared seedling in holes that are made for the plant.

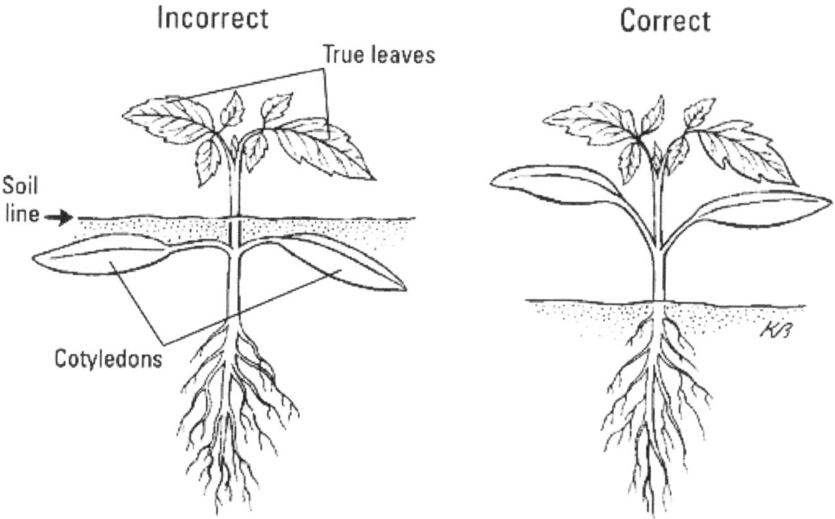

- ☆ Plant seedlings at the correct depth as tomatoes need deeper planting.
- ☆ Gardener should remove the entire 3 or 4 sets of leaves from the top before planting.
- ☆ After firming the soil around the roots with the hands, gardener should form a shallow soil basin around the base of transplant.
- ☆ The soil basin will work as moat around the seedling to hold water.
- ☆ In case of rains or more water, moisture stays in moat and drains to where the roots are located.
- ☆ As per the conditions, water the bed the same day or next day.
- ☆ If weather is dry or if soil is sandy, gardener should water the entire bed and when it's rainy or soil is wet, then gardener should wait for next day to give water.
- ☆ Gardener should keep the bed moist while the seedlings get established and begins to grow strongly.

1.20.4 Precautions to be taken while Transplanting Seeds

While transplanting seeds, the gardener should take care of the following steps:

- ☆ Water the seedlings before pulling.
- ☆ Pull the seedlings carefully without shaking off the media from the roots.
- ☆ Plant the seedlings immediately to prevent the roots from drying out.
- ☆ Gently press the soil around the roots.
- ☆ Water the seedlings at the soil surface with a gentle stream of water to settle the soil around the roots.

1.21 Marketing Survey and Feasibility

1.21.1 Introduction

The marketing of cut flowers is much unorganized. In many cities, flowers are brought to wholesale markets, which mostly operate in open yards. From here, such flowers are distributed to the local retail outlets which operate in open on-road sides, with different flowers arranged in large buckets. The packaging and transportation of flowers from farms to retail markets are very unscientific. The flowers, as per the selection get packed in gunny bags, bamboo baskets, simple cartons or are wrapped in old newspapers and are transported to respective markets by road, rail or by air.

1.20.2 Features

- ☆ Many flower growers are selling value-added products with high markups online.
- ☆ The flower growers are selling dried flower wreaths, bulbs, seeds, dried flower bundles and dried flowers for the wedding.
- ☆ Lavender growers are selling lavender sachets, lavender soap, dried lavender wreaths and bunches, lavender oil and lavender buds.
- ☆ Current markets for plant material produced in ornamental nursery operations include the general public, landscape contractors, independent or chain retailers, other production nurseries, brokers, government agencies or a combination of these.
- ☆ Production nurseries that sell to the general public may do so via an affiliated retail nursery or by mail order.
- ☆ Production nurseries selling to retail operations may further specialize in sales to independent retail garden centers, chain retail garden centers or big stores.
- ☆ Sales to retail garden centers typically call for a more diverse group of plant species and higher quality, higher-priced products in a greater selection of cultivars and sizes.
- ☆ Such market type usually involves many different customers in various locations needing different products at different times.

- A production nursery servicing retail garden centers has its own trucks for delivering plants.

1.20.3 Market Considerations

- One of the first rules of business is to know the potential markets.
- In nursery business, one should determine who the customer is, where its available markets are in relation to its customer base and what products it might produce for those markets.
- It is best to first identify a market and then develop a nursery plan to produce and sell to that market.
- Many nurseries are started based on the owner's personal favorite plant's sale or that a nursery can be a method of getting rid of excess plant materials from personal plant collections.
- If a market for the product is not first identified and properly exploited, then such a nursery will not be so successful in the long term.

1.20.4 Factors

A nursery is often defined by its method of production, its market, crops grown or sizes of plants sold. Crop selection is a critical component of nursery production, marketing, and sales. The types and diversity of crops and plant sizes have a great impact on nurseries' ability to market plant material to the appropriate target audience and realize maximum profitability. The types and sizes of crops grown may influence the following:

- Nursery size: Liner nurseries take up much less space than large-container or B and B tree production; as a result, the property needed for a liner nursery is much smaller.
- Production requirements: Cold-sensitive crops and container-grown crops may require greenhouses or winter protection, whereas cold-hardy plants or field-grown trees may not require any specialized structures.
- Market: Liners are sold to other nurseries, whereas larger plants, also called "finished," maybe destined for retail sales or landscape contractors.
- Location: Nursery crops and markets often influence nursery location. A nursery selling to garden centers is best located geographically close to them, typically in suburban or urban areas, whereas a liner nursery does not have this requirement.
- Production Costs: Production costs are important considerations. Supplies such as plants, containers, substrate, fertilizer, chemicals and other items represent about 26 per cent of operating expenses. It is advantageous to locate a nursery near suppliers of these production components to reduce

shipping and other costs. An alternative is to locate a nursery near other nurseries to cooperate on group purchases to reduce shipping costs and obtain better prices on bulk purchases. In both of these cases, production costs will be reduced on a per-plant basis.

- Marketing Costs: The type and location of customers greatly affect marketing costs. Many retail sales need a color point of sale tags, a comprehensive marketing plan, and perhaps a branding program to maximize exposure, impact, and sales to the retail customer. Marketing costs associated with sales to landscapers and other nurseries typically are not burdened by these marketing expenses, though there are costs associated with trade shows, websites/catalogs and plant locator services. Nursery sales may be handled by in-house sales staff, brokers or combination of the two.

Chapter 2

Cultivation of Flower Crops

2.1 Types of Flowers for Cultivation

2.1.1 Introduction

Growing flowers is a dream for many people and a full-time work for floriculturist for commercial purposes. Growing flowers serve as interests of diverse people. The growing of flowers necessitates a lot of patience, love and care.

2.1.2 Types of Flowers

Flowers are classified into three categories:

- ☆ Annuals
- ☆ Perennials
- ☆ Biennials

Annuals grow from seeds and flower the current year and then die. Perennials establish themselves from seed the first year but do not usually flower until the following year. Such a category of flowers will re-grow and flower for many years to come. Biennials grow as small plants during the year they are planted while in the second year they will bloom and perish.

2.1.3 Selection of Flowers

Growing of flowers after careful selection will give the best quality flowers. For good flower selection, floriculturist has to consider certain important criteria's that are generally considered before deciding the type of flowers that needs to be grown, which includes knowledge about:

- ☆ Native flowers are always a good option to grow.

- The shelf life of flowers to be grown should be kept it mind.
- For commercial flower farm, floriculturist should select flowers with high commercial value.

Floriculturist while selecting flower type, should also have knowledge related to:

Annual Flowers

- Annual flowers complete the whole process from potting to blooming over the course of a year, as they can die at the end of the growing season and needs to be repotted at the beginning of every new season.
- Annuals grow fast, so if floriculturist feels the need for establishing a pretty garden, then it is the best option.
- Annuals need more care and cultivation as compared to perennials.
- Such flower type is required for the full lush garden but at the cost of extra work, as it can die at the end of the year.

Perennial Flowers

- Perennials are flowers that bloom every year and need not be repotted.
- Such flower type gives an edge over annual flowers where floriculturist doesn't have to repeat the whole process from potting, nurturing and finally blooming every year.
- Once such flower type gets potted, they can live for years.
- Such flower type is easier to pot and to cultivate as compared to annuals but, don't bloom in every season.
- Once potted, a perennial might take up a year to bloom flowers.
- However, gardeners often look forward to their favorite perennials blooming as per the type of season.

2.1.4 Flowers for Cultivation

Flowers bring smells and color as some are perennials while some are annuals. Some flowers are fragrant and some without smell while some have different colors. There are certain flowers that are hard to maintain, while others can be maintained easily.

Summer Season Flowers

Dahlia

- It is a beautiful perennial which is a source for many arts and cultures.
- It blooms every year in summer and needs plenty of the sun.
- It requires no shadings.
- It should not be kept directly in sunlight.

- It blooms for a full 120 days and is absolutely stunning to behold.
- It needs soil pH of around 6.5-7 with acidic soil.
- The soil should be well-drained and rich in organic matter.
- The use of bone-meal as fertilizer before planting will help to go a long way.
- They are tuberous and grow out of large tubers.
- It needs no water until the plant shows dryness as excess water can lead to the death of the plant.

Sunflower

- It is a little summer flower that attracts butterflies.
- It needs a lot of sunlight.
- It needs no fertilizer and can reseed themselves.
- It grows best in hot weather.
- It is an annual which blooms at the end of summer and needs less maintenance.

Calendula officinalis

- It is a summer flower which is known for soldiering by even early frosting.
- It provides a vibrant orange and yellow color.
- These annuals are edible and have several medicinal properties.
- Its petals are good for skin, and eating the leaves adds a bit of zest to salads as they are spicy.
- They can grow in partial shade and in full sun.
- It grows in 45-60 days of seeding.
- Ample liquid fertilizer is good for getting beautiful big specimens.
- The soil needs to be rich in compost and moisture.
- Due to toughness by the late fall season and their edible medicinal qualities, they are all-around good flowers to be in any area.

Eschscholzia California

- It is a perennial flower that comes with many varieties of colors.
- It needs low maintenance as many perennials and requires less watering and no fertilizer.
- It blooms from early summer to early fall.
- It can grow in any warm climate as long as there are ample sunshine and less watering.

- ☆ From sowing, seeds take up 15 days to germinate which makes it a quick and easy addition to the garden during summers.
- ☆ With abundant leaves, it provides lush foliage to offset the vibrant colors in the summer garden.

Fall Season Flowers

Chinese Bellflower

- ☆ It is also known as a balloon flower which is a blue bloom.
- ☆ It is perfect for receding hot summer and cool fall breeze.
- ☆ It grows in partial sun or shade and is favorite among children.
- ☆ Such flower type with bulb forms before blooming is fun to pop open.
- ☆ It requires a good bit of moisture and needs to be maintained accordingly.
- ☆ Such peace-loving perennials do not like to be disturbed as they have quite extensive trap root and grow best when left alone, once potted.

Michaelmas Daisy

- ☆ It is an autumn aster having little pink-lavender flowers.
- ☆ It provides excellent compliments to any ornamental grass.
- ☆ It grows quickly in ample sunlight and needs well-drained soil.
- ☆ It needs moisture but not too much.
- ☆ It can grow best when the soil is only temporarily moist and dry by nightfall.
- ☆ It needs fertilization.
- ☆ Being perennials, it can bloom within the first year of planting.

Caryopteris

- ☆ It is known for its blue beauty.
- ☆ Even it needs fast-draining soil. It tends to be very hard.
- ☆ The flower type is one of few that comes close to that clarity blue.
- ☆ It produces a beautiful lavender fragrance which is often under appreciated in comparison to its appearance.
- ☆ Growing all the way till winter starts, such blue flowers are the pride of any self-respecting fall gardener.

Heliopsis

- ☆ It is sun-like blossoms that looks just like sunflowers.
- ☆ Such flower type needs little or no maintenance.
- ☆ Once seeded, it can grow quickly in fast-draining soil like most perennials.

- They are also very good at reseeding, so there might be a need for curbing their growth if they get too abundant.
- Such little false sunflowers are a perfect complement to blue and purple that usually pervades fall gardens.
- It attracts butterflies easily, so the garden gets even more color splashed in.

Winter Season Flowers

Cineraria

- It is mostly found in gardens and shows a vibrant dash of pink, blue, red, orange and bi-colored varieties.
- Such flower type takes 10-15 days for the seeds to germinate.
- It requires a lot of sunlight to germinate.
- They are used as outdoor annuals as they are very fragile and need to be repotted once they die after blooming.
- It is a high-maintenance flower.

Daisy

- Daisies are popular winter flowers.
- They prefer shade to sunlight, but a little bit of sunlight is necessary.
- These perennials need less maintenance.
- They do not need much watering or fertilizer.
- They can be planted in poor soil and prefer low-nutrient soil in comparison to high organic soil.
- Due to high adaptability and low-maintenance, they can pop up in the garden under certain conditions.

Spring Season Flowers

Pansy

- It is an annual flower that serves as the choice for the first bloom of spring.
- It comes with blue, purple, red, orange, pink and yellow-colored petals.
- Pansies like the sun but not heat and are kept where there is plenty of sunlight during cool weather.
- They need to be planted 6-8 weeks prior to blooming, so winter sowing of seeds is required.
- It requires lots of water and good fertilizer.
- They attract butterflies as they are filled to the brim with nectar which adds a bit of flair to the garden.

Trillium

- ☆ They are special flowers.
- ☆ The flower type has groupings of three petals, three leaves, and three sepals.
- ☆ They are also sometimes referred to as "Trinity Flower."
- ☆ They come in white, yellow, red and other colors depending as per the region.
- ☆ They are mostly found on forest floors as they are not very hard to grow.
- ☆ They require partial shade, moist well-drained soil, and rich organic matter.
- ☆ They are perennials that bloom in spring and once potted are there to stay.

Bloodroot

- ☆ It is a beautiful white flower that has no leaves.
- ☆ It has a dignified elegance which only a few other flowers have.
- ☆ They are low-maintenance flowers that once potted need very little care.
- ☆ It has deep-red sap that is produced by its stem and roots.
- ☆ Such plant type is used for making dyes of pink, orange, and red.
- ☆ It has been used medicinally in some places over the centuries, and that just adds to its charm.
- ☆ Such perennials need moist soil and regular watering at least twice a week.
- ☆ They don't need much sun and can be placed in the shaded parts of the garden.
- ☆ It needs fertilizer and can be self-propagating.

Lilac

- ☆ These are beautiful flowers that are popular spring addition to gardens.
- ☆ They are known for their delightful fragrance.
- ☆ It does well under the full sun but can do with the partial sun as well.
- ☆ It requires neutral to slightly alkaline soil as far as pH balance is concerned.
- ☆ It is available in blue, purple, red, white, yellow and pink colors.
- ☆ The color of purple lilacs is the origin for color lilac, which is a variant of purple.
- ☆ They are hard and easy to maintain.
- ☆ They can grow pretty tall with tallest ones reaching 15 feet.
- ☆ They need at least 6 hours of sunlight so make sure to not put them under shade.
- ☆ Pot these precious flowers and enjoy the haunting fragrance that they emanate.

2.2 Estimate Nutrient Requirement for Flower Crop

2.2.1 Introduction

Nutrient management is critical in order to increase/maintain crop yields on a single part of agricultural land. To meet crop needs throughout the growing season, soil fertility must be consistently high. There are around 14 nutrients that are supplied to crops from soil and fertilizer sources out of which, six are macronutrients while the rest are micronutrients. Macronutrients are used in large amounts by plant/flower while deficiencies of micronutrients are equally damaging to yield and profitability.

Every flower has some basic needs; the most important of all is that it needs to bloom, just like all living things in general, albeit in a symbolic way. It needs nutrition and water of course, but most of all it needs sunlight. The soil is another crucial thing that a plant needs. It is where it gets the rest of the energy from. The soil cannot be too sandy, but too sticky is also not good.

The six main nutrients essentials to healthy and fruitful plant development are carbon, hydrogen, oxygen, nitrogen, phosphorus, and potassium. The first three are in air and water, and the last three must come from the soil.

The nutrient requirement in flower crops will depend on:

- ☆ Target yield
- ☆ Soil nutrient supply
- ☆ Type of fertilizer and recovery
- ☆ Timing of maturity

2.2.2 Crop Nutrients

Nitrogen, phosphorous and potassium are primary nutrients that are supplied by inorganic fertilizers. Nutrients are also supplied by other products and processes such as organic manures, plants residues and biological nitrogen fixation.

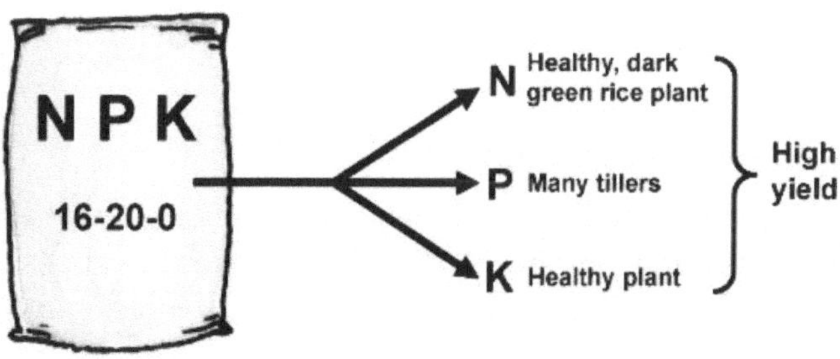

Practical nutrient needs for certain crops comes from regionally specific guidelines which depends on crop and environmental factors like soil type and weather conditions and amount of nutrient being applied.

Essential Plant Nutrients

Element	Chemical symbol	Chemical forms absorbed by crop plants
Primary Nutrients		
Nitrogen	N	NO_3^-, NH_4^+
Potassium	K	K^+
Phosphorus	P	$H_2PO_4^-$, HPO_4^{2-}. PO_4^{3-}
Secondary Nutrients		
Calicium	Ca	Ca^{2+}
Magnesium	Mg	Mg^{2+}
Sulfur	S	SO_4^{2-}
Micronutrients		
Boron	B	BO_3^{3-}
Chlorine	Cl	Cl^-
Copper	Cu	Cu^+, Cu^{2+}
Iron	Fe	Fe^{3+}
Manganese	Mn	Mn^{2+}
Molybdenum	Mo	MoO_4^{2-}
Nickel	Ni	Ni^{2+}
Zinc	Zn	Zn^{2+}

Calcium, magnesium, and sulfurs are secondary macronutrients or secondary nutrients as they are required in the lesser amount as compared to primary nutrients. These elements are equally important for plant growth and nutrition. Often, adequate amounts of these nutrients can be found in the soil and supplied to crops from chemical weathering or atmospheric deposition.

Boron, zinc, manganese, iron, copper, molybdenum, nickel and chlorine are micronutrients that are essential for plant growth which is used in small amounts by flower crops. The deficiency in any of these elements results in the potential to cause a decrease in crop quality or yield while excess availability of such nutrients may cause toxicity.

2.2.3 Nutrient Deficiency Sign

If flower plants are not thriving or producing flowers as expected but are getting enough sun and water and are not infected by disease or insect pests, then it may result in nutrient deficiency or imbalance. Nutrient deficiency or imbalance can be identified by:

- Chlorosis or yellowing of plants and leaves
- Necrosis which is the death of plant tissue

- Accumulation of anthocyanin, which results in purple or reddish plant parts
- Lack of new growth
- Stunted plant growth

For flowering plants, nutrient deficiency and imbalances may present as a lack of buds or poorly developed fruit.

- If gardener/floriculturist suspect a nutrient imbalance, in such a case, he/she should do soil analysis.
- Soil analysis is a soil test that provides a nutrient profile and pH reading.
- If in the test, a pH is too high or too low, then nutrients available in the soil will not be properly absorbed by plant roots.
- Fertilizers can also be used to correct the soil imbalance which is a common source of nitrogen and potassium having calcium ammonium nitrate, urea, potassium chloride, and potassium sulfate.
- Phosphorus sources include superphosphate or bone meal.
- Organic soil amendments, such as livestock manure, plant compost and natural sources of nutrients, can also be used to remedy a mineral imbalance, ensuring healthy growth of flowering and fruiting plants.

2.2.4 Determining Nutrient Needs

- Soil testing is an important part which will help in determining such nutrients that are already available and those which are in limited supply.
- Deficiencies of nutrients will impact the crop quality or yield.
- Knowing about such nutrients will help in determining the amount and type of fertilizer required.
- Soil testing depends on appropriate soil sampling on an individual field basis.
- Proper soil sampling guidelines can be accessed online and by extension resources.
- Plant tissue analysis will be used as a method for measuring nutrient levels of the plant during their growth.
- The analysis of plant tissue is a good tool for growers who can understand the effectiveness of fertilization practices.
- The tissue analysis is also good for diagnosing certain production problems.
- Such tissue analysis is also good as the samples that are submitted and its guidelines can also be obtained easily.

2.3 Crop Spacing and Border Crops

2.3.1 Crop Spacing

Introduction

Crop spacing in flower plants is all about the number of flower crops planted in a unit area. It is about the distance between one flower plant with others. Agriculturalists/Floweriest have advice about plant spacing with respect to particular crops that are based on the research, which shows the idea to maximize flower crop yields.

Gardeners/Farmers want to fully use their land for getting the best yields and failed to follow recommended flower crop guidelines, and lead to more money loss, time and soil fertility. It is advisable for gardeners/farmers to follow recommended guidelines so that different annual crops should be well spaced to achieve the best results. It is also known that summer annuals can be planted farther apart as they have all summer to grow.

The flower plant spacing had significant variation in the growth and yield components which exclude pericarp thickness. Number of branches per plant, number of leaves per plant, stem girth, number of fruits per plant, days to the first harvest, fruit length, individual fruit weight, yield per plant were found to be significantly increased with increasing of plant spacing but plant height at different stages, number of fruits per plot, days to 50 per cent flowering, fruit breadth, yield per plot and yield per hectare were found to be significantly increased with the decreasing plant spacing.

Space Requirements

- ✰ When plants are just seedlings or cuttings they can be grown in close proximity, but as they grow their need for space will increase and they will need to be moved or thinned out.
- ✰ In farms/gardens, spacing can be created by removing the smaller or weaker plants and making space for stronger and more aggressive plants to grow.
- ✰ With more living space available for remaining flower plants, they can efficiently use the available area for healthy growth as it more area results in more airflow which helps in preventing mildew and more light for photosynthesis.

Features

- ✰ To have little less maintenance, the annual flowers should be spaced around 8-12" on center planting pattern to give space for plants to spread and look better.
- ✰ When flowers are planted close together they need more fertilizer, as there are many hungry roots in a small area, since soil can't properly nurture them all.

- ☆ If the flower plant is too close together, then the plants get stressed and are prone to diseases.
- ☆ If the air inside the crop can't properly circulate, then plants will not be able to dry out due to watering, which allows the fungus to sets in the roots which makes the plant weakened from stress, allowing insects to move in.

Benefits

- ☆ It is always good for gardener/flouriest to follow the recommended crop spacing guidelines.
- ☆ Spacing of plants is an important aspect that needs to be carefully considered and properly addressed for successful hydroponic system.
- ☆ In the event that plants are grown too close to one another, they will likely receive less than the required amount of light.
- ☆ Overcrowding of crops reduces yields and lower the quality of fruits produced due to competition for light and soil nutrients.
- ☆ If crops are planted too close to one another, it may be hard for gardener/ farmer to walk about in the garden weeding, spraying pesticides or doing crop inspection.
- ☆ When a farmer practices intercropping, it is particularly important to pay attention to spacing.

Recommendation

- ☆ The farmer should constantly monitor the flower crop's growth and should do thinning where possible to ensure correct and uniform plant population.
- ☆ To ease weeding, harvesting, spraying and general crop inspection, gardener/farmer is best advised to plant the flower crops in the rows.

2.3.2 Border Crops

Introduction

The plants in border beds and along edging helps in defining areas of outdoor space and offers transition from one space to other. Borders at patios, walkways or driveways soften landscaping elements and add visual interest to the space. Although there are many plants that can be used in borders and edging, designing the space effectively is the key to showcasing each plant and creating an attractive border bed. Flower garden border pulls out all the stops with multiple components to keep the viewer interested. In addition to perennials, it exhibits:

- ✰ Annual plants
- ✰ Pachysandra
- ✰ Evergreen shrubs

Types

Spring Border

In springs, yellow alyssum such as creeping phlox is used as flowering ground to cover plant. It contrasts strikingly with grape hyacinths not only in color but also in another respect: It is a bad-smelling flower and if flowers are not required for fragrance, then such flower will give pleasant appearance with its color.

Summer Border

Flower for creating borders in summer include flower with red, white and blue colours. Such patriotic displays are usually composed, at least in part, of annual flowers. Annuals are inexpensive bedding plants which allow to plant an impressive display without breaking the bank.

In such patriotic border, red color is achieved with geraniums, white is achieved with annual alyssum, and blue is achieved by blue ageratum, by using:

- ☆ Red salvia (red)
- ☆ Shasta daisy (white)
- ☆ Victoria Blue salvia (blue)

Autumn Border

Flowers used in autumn for creating borders include autumn joy sedum which is late-bloomer which is ideal for autumn displays. With its long-blooming perennial plantation, such flowers assure continuous garden color from late summer until first hard frost.

In the garden, Autumn Joy sedum is complemented by Silver Dust dusty miller, among other plants. The dusty miller is a common choice when silver foliage is recommended. The brightness of its frosty, white fronds leavens a scheme that is otherwise dominated by the dusky-pink tones of the Autumn Joy.

Variety of Borders

Ground Covers

- Ground covers are often used in front of a border due to their short stature which adds color and interest without covering taller plants in the middle or back of the border.
- Plants placed in front of the border will help in creating a transition from the walkway or grass to planting bed.
- Using variety of ground covers will ensure for long-lasting colours during the season.
- With different size, growth habit and texture of plants at the border, the front edge helps in keeping the bed looking attractive.
- The front of the border is also a good spot to plant smaller bulb plants and herbs.

Midsize Plants

- Planting midsize perennials, plants between 1 and 3 feet tall, in the middle of border bed adds height to the bed which allow plants in back of the bed to be seen.
- Taller perennials and smaller evergreen or deciduous shrubs add multi-season color to the middle of the border.

Tall Plants, Shrubs and Trees

- The plants at back of the border should be the tallest which should allow them to be seen over the plants in front and middle of the border and offers attractive backdrop for plants that are in front.
- Selecting plants taller than 3 feet for back of the border includes tall shrubs and trees like white forsythia, which is a relative of yellow-flowering forsythia that blooms with white flowers in spring.
- The deciduous shrub grows 3 to 6 feet tall.

Other Options

- Annuals are good for making borders, as it allow gardeners to easily change the appearance of the border as per the season.
- Annuals like wishbone flower or sweet alyssum bloom from spring through fall while ornamental grasses like Ebony Knight mondo grass have purple-black foliage which adds contrast to surrounding green foliage.

2.4 Time of Planting Flower Crop

2.4.1 Planting

The gardener should understand flower planting process which can be either on ground or in a pot:

- ☆ After deciding the point of planting, a hole with a diameter similar to that of the plant's root ball must be dug.
- ☆ If the soil is not of good quality, compost can be added to the bottom of the hole.
- ☆ The root ball is removed from the pot and placed in the hole.
- ☆ The hole must be filled with soil and lightly pressed, so the root ball makes good contact with the soil.
- ☆ The new plant should be watered, and stakes can be used if necessary.
- ☆ The stakes must be tied loosely to the trunk and not dig into the bark or tighten around the tree.

If the plant is placed in a pot, the procedure is even simpler:

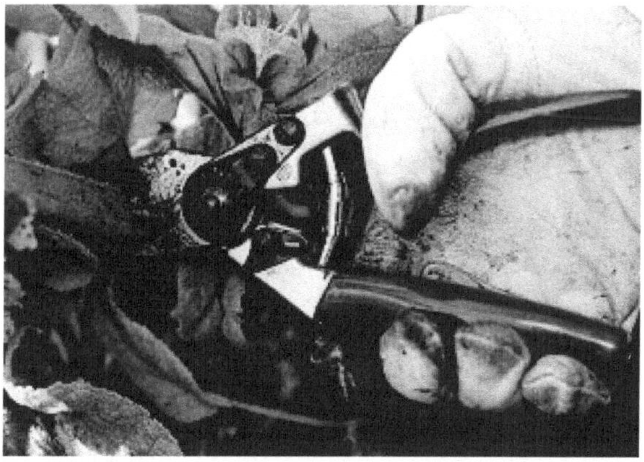

- ☆ Some broken tiles or small stones should be placed to the bottom of the pot, to keep the holes open for water to run through.
- ☆ Some soil must be placed.
- ☆ The plastic pot from the plant is removed and placed to the pot in use.
- ☆ Soil should be added for filling the gaps and covering the root ball of the plant.
- ☆ Lightly press and then water.

2.4.2 Time of Application

- ☆ Spring is best time to fertilize when outdoor plants start growing again.

- ☆ There is time when fertilizers release nutrients, so a gardener should know when to re-apply.
- ☆ Many granular fertilizer is slow-release which can be applied only once or twice a year for perennial and mixed borders.
- ☆ Lawns can be fertilized 2-4 times a year as per general condition, as fertilizer spikes are put in ground once in spring.

- ☆ Anything mixed with water is easily used by plants so it can be applied more often.
- ☆ Grass needs fall feeding, so it is better to stop fertilizing after September and rose plant by August for natural cycle of dormancy instead of forcing growth.

2.5 Seed Rate

2.5.1 Introduction

Seeds are the base of agriculture. Technology has modernized much of farming's day-to-day operations, but without steady supply of high-quality seed, yields and crop quality would be greatly decreased. Seed quality also plays an important role in production of agronomic and horticultural crops. Characteristics such as trueness to variety, germination percentage, purity, vigor, and appearance are important to gardeners/farmers in planting crops and to homeowners in establishing their lawns and gardens.

Numbers of plants in a crop depends on seed rate, seed viability, percentage seedling emergence and plant survival. However, plant numbers can often vary widely without appreciably affecting yield in irrigated crops as plants produce tillers, which themselves produce leaves, spikes and grain. There are many sizes and shapes of seed which are from micro-miniature orchid seed, as small as particle of dust, to hat of gargantuan double coconut seed, which is 1 feet long and weighs many pounds. Shapes of seed vary from simple, round tobacco seed to complex, aerodynamic, winged maple seed. Some seeds are hairy like cotton or thistle while others are like grass, have long spikes while some have thorns.

Seed Rates in Marigold

Seed rate for marigold flower varies from 2.0 – 2.5 kg per hectare and takes about 5-7 days for germination. To raise seedlings, seeds can be sown in pots, seed boxes or on flat or raised nursery beds in lines with shallow furrows. During preparation of nursery beds, 8-10 Kg of well decomposed farm yard manure per 1 sq.m bed is mixed with the soil.

2.5.2 Factors to Consider for Seed Rate

The required number of plants per unit area is decided by calculating the seed rate which depends on:

- Plant spacing
- Plant population
- Test weight
- Germination percentage
- Weight of seed
- Plant type; tall or short

Percent Germination

In addition to the number of seed per kilogram, gardeners/growers should also consider percent germination which is proportion of seed that germinates while determining the weight of seed required per acre. As of hereditary and/or seed quality factors, only a part of seed (70 per cent -90 per cent) in a given lot will germinate. Percent germination is determined by subjecting samples of specific seed lots to conditions conducive to germination.

The percent germination calculated should be shown on seed label. It is the duty of grower to check the label in order to confirm the percent of seed expected to germinate. If there is any doubt about the percent germination, a germination test should be done before planting the seed.

Weight of Seed

The number of seeds needed for weighing a pound, is known as seed count, which depends on weight of individual seeds. The weight of each seed for a given species is obtained by seed size and specific gravity of seed. Larger seed are heavier, which takes less to weigh a pound. Variation in seed size can have a two-fold effect on recommended seeding rates.

2.5.3 Calculating Seed Rate

To calculate seed rate, the following formula is used:

$$\text{Seed rate (kg/ha)} = \frac{\text{Plant population (per ha) x No. of seeds/hill x Test weight (g) x 100}}{\text{1000 x 1000 x Germination percentage (per cent)}}$$

2.6 Integrated Nutrient Management

2.6.1 Introduction

Integrated Nutrient Management involves maintenance of soil fertility and supply of plant nutrient at optimum level for sustaining required productivity by optimizing benefits from all possible sources of organic, inorganic and biological components in an integrated manner.

2.6.2 Concept

Integrated Nutrient Management uses following major sub-concepts like:

1. Plant nutrients stored in the soil.
2. Plant nutrients, those present in the crop residues, organic manure and domestic wastes.
3. Plant nutrients purchased or obtained from outside the farm.
4. Plant nutrient looses *e.g.* those removed from field in crop harvest and lost from soil by volatilization (ammonia and nitrogen oxide gases) and leaching (nitrate, sulphate *etc.*)
5. Plant nutrient outputs *e.g.* nutrient uptake by the crops at harvest time.

2.6.3 Goal

- ☆ To ensure a productive and sustainable agriculture.
- ☆ To reduce expenditure on costs of purchased inputs by using farm manure and crop residue, *etc.*
- ☆ To utilize the potential benefits of green manures, leguminous crops and bio-fertilizers.
- ☆ To prevent degradation of environment.

☆ To meet social and economic aspiration of gardeners/farmers without harming the natural resource base of agricultural production.

2.6.4 Components

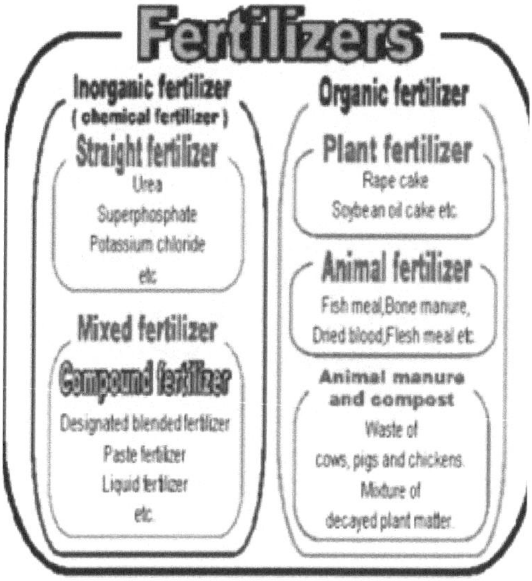

☆ Integration of soil fertility restoring crops like green manures, legumes *etc.*

Green Manure

☆ Recycling of crop residues

Composts

☆ Use of organic manures like FYM, compost, vermicompost, biogas, slurry, poultry manure, bio-compost, press mud cakes, phosphocompost

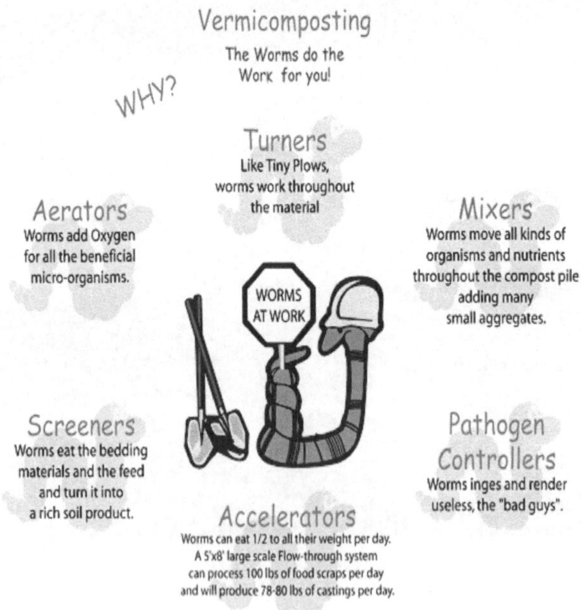

Vermicompost

☆ Utilization of Bio fertilizers

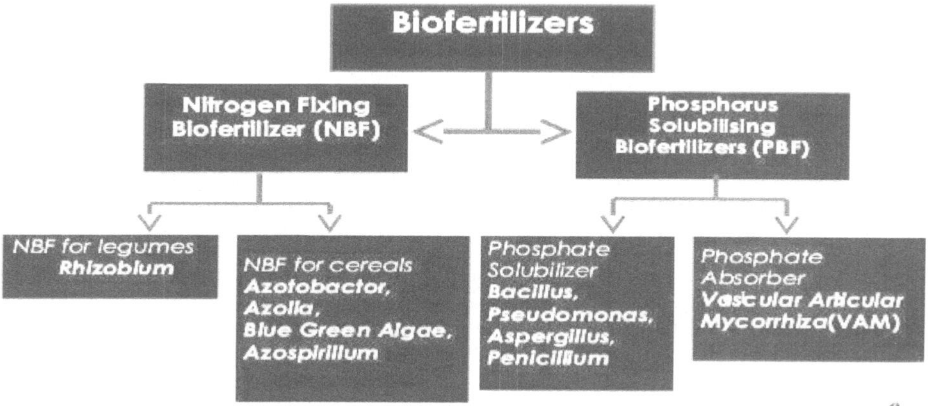

☆ Efficient genotypes

☆ Balanced use of fertilizer nutrients as per the requirement and target yields.

2.6.5 Need

The increasing use of chemical fertilizers increases the production of food and fiber which causes concern for the following reasons:

☆ Soils that receive plant nutrients only by chemical fertilizers are showing declining productivity in spite of being supplied with sufficient nutrients.

☆ The decline in productivity is attributed to the appearance of deficiency in Secondary and micronutrients.

☆ The physical condition of the soil is deteriorated as a result of the long-term use of chemical fertilizers, especially the nitrogenous ones.

☆ It aggravates the problem of poor fertilizer nitrogen use efficiency.

☆ Excess nitrogen use leads to groundwater and environmental pollution apart from destroying the ozone layer by N2O production.

2.6.6 Advantages

☆ Enhancing the availability of applied and native soil nutrients.

☆ Synchronizing nutrient demand with native supply from native and applied sources.

☆ Providing balanced nutrition to crops.

☆ Improves and sustain the Physical, Chemical and biological functioning of the soil.

☆ Minimizing deterioration of soil, water and ecosystem by promoting carbon sequestration.

- ✰ Reducing nutrient losses to ground surface water bodies and atmosphere.
- ✰ Minimizing antagonistic effects from hidden deficiency and nutrient imbalance.

2.6.7 Principles

The garden/farm profitability and water quality can be improved by efficient nutrient use. Manure and biosolids need to consider as valuable fertilizer materials which should be managed in the same way as commercial fertilizers. Soil testing is important for the development of nutrient application rates. The gardener should know the general recommended principles of nutrient management, which involves:

Nutrient Recommendations

Nitrogen

- ✰ Nitrogen recommendations for many crops are based on yield goals for those crops.
- ✰ It is important to establish realistic yield goals for each field based upon historical yield data.
- ✰ Recommended application rates for nitrogen should not be exceeded.
- ✰ The use of the Pre-Sidedress Nitrogen Test is recommended in early summer after forage legumes or manure and biosolids applications to corn in order to show. if extra nitrogen is needed.
- ✰ Residual values for nitrogen from legumes in rotation or previous applications of manure or sludge are deducted from total nitrogen recommendations.
- ✰ Growing winter cover crops is an effective practice for reducing nitrate losses from cropland, during the time of year when leaching potential is high.

Phosphorus and other Nutrients

- ✰ Recommendations for phosphorus, potassium, and micronutrients, depend on soil test values, yield goals, and crop rotation.
- ✰ If soil test levels are high, extra nutrients other than in-row, starter fertilizer are not required for many crops/plants.
- ✰ Soil pH influences nutrient availability, particularly phosphorus.
- ✰ Soil pH should be adjusted to the level recommended for crop/plant to be grown.

Application of Nutrient Sources

Proper timing of nutrient applications is important

- Apply nutrient sources as close to planting or nutrient demand as possible so that nutrients are taken up by plants quickly and not allowed to runoff into surface water or leach into groundwater.
- Avoid application of nutrient sources to frozen ground and during periods of high potential for leaching and runoff.
- Application in late fall or winter of any nitrogen source for a spring-planted crop should be avoided if possible.
- Avoid the application of nutrient sources to sensitive areas, wetlands, sinkholes, and steep slopes.
- Calibrate nutrient application equipment accurately to insure that recommended rates are applied.
- Accurate and uniform applications of nutrients are necessary to maximize the nutrient potential of the fertilizer materials.

Manure Applications Recommendations

Testing

- Manures vary tremendously in nutrient content as per animal species, rations, and storage conditions.
- The nutrient content of manure can be determined by laboratory testing.
- Whenever possible, manure should be sampled at least 6 weeks before planned application to allow time for analysis and plan development.
- A consistent baseline for nutrient content needs to be laid down and based on analyses taken at least twice a year till the uniform value is confirmed, and then every second year thereafter to verify its consistency.
- If significant changes occur like feed, management, animals or storage, then new samples should be collected for nutrient analysis.

Application of Manure

- Nutrient applications should be made at times of year that will minimize N and P losses to water and N volatilization loss to the atmosphere.
- Crop utilization of nutrients in manure and biosolids is maximized if these materials are applied in synchrony with periods of crop uptake.
- Storage of manure may be necessary to facilitate the appropriate timing of nutrient applications.
- Nitrogen-based applications of manure will cause phosphorus soil test levels to increase over time.
- Application recommendations for daily haul operations include consideration of slope, crop and vegetative cover.

Storage Capacity

- Optimal utilization of nutrients in manure and other nutrient sources is difficult without the ability to store manure for part of the year.
- Improving storage capacity will minimize the potential for nutrient loss or runoff and improves the possibility of proper timing of manure applications.

Erosion and Runoff Control

- Best Management Practices should be used to minimize soil erosion and runoff, which carries nutrients to surface waters.
- Best Management Practices around the barnyard area may need to be updated based on current regulations to reduce the likelihood of nutrient loss from the area.
- The addition of any P-bearing material to fields whose P soil test levels are greater than or equal to FIV 150 will require evaluation of the risk of P movement.

2.7 Method of Application of Fertilizers

2.7.1 Fertilizers

Fertilizers are chemical compounds applied to promote plant growth. They are applied either to the soil or by foliar feeding.

Soil composition changes, if different fertilizer is added to it. There are bulky organic fertilizers like cow manure, bat guano, bone meal, organic compost, and green manure crops. Apart from it, there is chemical fertilizer, which is known as inorganic fertilizer that is made from different formulations to suit a variety of specified uses.

Compared to organic compost, chemical or inorganic fertilizers also have certain advantages of being less bulky. Being less bulky, chemical fertilizer can be easily transported, both overland and from soil to plants itself, as they made to be available to the plant relatively faster when incorporated as part of plant-food constituents.

Chemical fertilizer comes in either granular or powder form in bags and boxes, or in liquid formulations in bottles. The different types of chemical fertilizers are classified as per principal elements, Nitrogen (N), Phosphorous (P) and Potassium (K), which may, therefore, included in more than one group.

Advantages of Organic Fertilizer

- It provides increased physical and biological storage mechanisms to soils, mitigating risks of over-fertilization.
- Such fertilizer nutrient content, solubility and nutrient release rates are much lower than inorganic fertilizers.

- With over 140-day period and after 7 leachings, these fertilizers release 25 per cent and 60 per cent of their nitrogen content, controlled-release fertilizers had a relatively constant rate of release while soluble fertilizer releases nitrogen content at first leaching

Disadvantages of Organic Fertilizer

- It is difficult to chemically distinguish between urea of biological origin and those produced synthetically.
- Like inorganic fertilizers, it is possible to over-apply organic fertilizers and, as a result, a water-soluble soil test should be conducted to determine optimum application rates.
- The release of nutrients happens quite suddenly as per the type of fertilizer used.
- These fertilizers from treated sewage, composts and sources can be quiet variables from one batch to next.
- Unless each batch is tested the amounts of nutrient applied are not precisely known.

Hazard of Over-fertilization

- Over-fertilization of the important nutrients can be detrimental as under fertilization.
- Fertilizer burn can occur if too much fertilizer is applied, resulting in drying of roots and damage or death of the plant.
- Organic fertilizers and specially-formulated inorganic fertilizers classified as slow-release fertilizers cannot cause nitrogen burn.
- If excess nitrogen is present, some plants can exude the excess through their leaves in a process called gestation.

2.7.2 When to Fertilize

Deciding when to fertilize is not as difficult as it might seem.

- Fertilizing is normally not necessary the first year after planting
- Never fertilize a plant in the late summer or early fall
- Use organic fertilizers when possible
- The overabundance of nutrients is as harmful as a deficiency
- A soil test is the best measure of nutrient status

Plant clues that soil fertilization may be necessary:

- Poor, stunted growth
- Chlorosis (leaves looking yellow or pale green)
- Lack of flowering and fruiting

2.7.3 Methods

The methods of applying fertilizers depend on the nature of flower plants, their nutrient needs, and soil. Gardener applies fertilizer to flower crops using the following methods:

Deep Soil Application

- ☆ Organic manures are kept on the surface of the soil and are incorporated in soil by using plow or rototiller before planting.
- ☆ Fertilizer can also be applied in bands at the bottom of plow furrows or broadcast or spread on top of plowed soil which is worked in soil using harrow before planting.

Broadcasting

- ☆ Gardener/farmer can use a walk-behind drop spreader or hand-held spreader with a crank to scatter granular or bulk fertilizer on the ground.
- ☆ Such a method is good for flowerbeds, lawns, trees and vegetable gardens.
- ☆ In this, fertilizer can be broadcast on the ground before planting and then tilled or watered into the soil.
- ☆ Top-dressing shows broadcasting on growing plants and water in the soil.

Banding

- ☆ To fertilize row of flowers/vegetables or to get plants started, the fertilizer is kept 2 inches to the side and 2 inches deeper than seed furrow.
- ☆ If irrigate with furrows, keep the band of fertilizer between irrigation and seed furrows.
- ☆ Keep bands of fertilizer on each side of the furrow.
- ☆ If the gardener needs to irrigate with drip hose, then he/she has to place the fertilizer under the emitter or keep fertilizer on one side of seedling or plant mid-way by its growing period which is known as a side dressing.

Liquid Application

- ☆ Water-soluble fertilizer can also be applied to flower crops using sprayers, sprinklers or can furrow irrigation.
- ☆ Small amounts of liquid fertilizer are applied to young plants at the time of transplanting.
- ☆ Many plants need micronutrients like zinc and iron which cannot be obtained from soil, where leaves can absorb them.
- ☆ Water-soluble fertilizer is sprayed on flowers/leaves when they first show signs of nutrient deficiency or in periods of drought when soil is dry to absorb nutrients.

- ☆ On using a strong solution on flower/plants, there are possibilities that the flower leaves get burn or scorch.
- ☆ Gardener/applicator should only apply a small amount of nutrients in a single spray.
- ☆ It is a costly method as fertilizer can be combined with pesticides to control insects or disease.

2.8 Specialized Practices such as Pruning, Pinching, Disbudding, Netting *etc.*

2.8.1 Pruning Practice

Pruning is an art and science of cutting away of portion of plant to improve its shape, to influence its growth, flowering and fruitfulness and to improve the quality of product. It is done to divert a part of plant energy from one part to another part of plant.

Features

- It is originated by the observation that simple disposition of branches in particular position is not sufficient to achieve crop but a definite fruit exists on the plants towards which if sap flows as to divert to plant can bear as abundantly and superior quality.
- Pruning has an effect on the function of the plant as it influences cropping of plants highly.
- It determines the capacity of the plant to produce fruit.
- With this, a gardener can actually drive the flow of sap towards the fruiting area on plant and force the plant for the better quality of fruits.
- It can also be done to eradicate pests and various types of fungi.
- In nurseries, it is common practice to prune plants before harvesting or transplanting as it increases both the quality and quantity of flowers.
- It is also used as a preventive measure to make space for any new seedling to grow.
- Such a method aims to reduce the size of trees and complexity of the classification model.

Types of Pruning

- Thinning out
- Heading Back
- Bulk Pruning
- Thin wood Pruning

Objects of Pruning

- Training of young trees
- Maintenance of grown-up trees *i.e.* to maintain the health of bearing plant
- Bringing vigor in old trees

Effect of Pruning

- It increases new vegetative growth
- In young trees, flowering will be delayed
- In old trees, there will be new vigorous vegetative growth which bears fruit
- It reduces bearing surface as a result of which, trees remain dwarf which is compensated by accommodating more number of dwarf trees
- Improvement in size, color, and quality of fruits

Principles of Pruning

- ☆ Young trees are pruned to train it to get the desired shape.
- ☆ In old trees, the light heading back is done to stimulate the flowering
- ☆ In bearing trees, a light pruning is done to stimulate fresh growth.
- ☆ In old trees, heavy pruning is done to restore vigorous

Systems of Pruning

- ☆ Heading back: In this, only top branches are headed back or cut off.
- ☆ Thinning out: In this, branches or part of plant/tree is completely removed.
- ☆ Dehorning: It involves cutting of main limbs or thick major branches
- ☆ Bulk pruning: It involves heavy pruning all over the tree.

Rules of Pruning

- ☆ Never leave a stub as far as possible
- ☆ Minimum cut surface
- ☆ Start cutting from the lower end first, leave halfway or even less and then cut from the top
- ☆ Keep the cut surface clean and smooth
- ☆ Protect the wound with Bordeaux paste.

Pruning of Fruit Plant/trees

Fruit trees are pruned for a number of reasons including:

- ☆ Architecture and size: Fruit trees are pruned to establish the basic tree architecture and to limit the tree size.
- ☆ Sunlight interception: Pruning opens spaces in the tree canopy for sunlight to penetrate and support leaf photosynthesis and the production of carbohydrates and energy that drive the growth of roots, shoots, and fruits. Without adequate sunlight penetration during the growing season, flower bud production can be reduced with detrimental effects on cropping in the future.
- ☆ Air circulation: Pruning can reduce branch crowding, reduce the amount of foliage inside the canopy and enable better air circulation that reduces humidity and allows for better coverage of protectant products (*e.g.*, pesticides).
- ☆ Strength: Branch angles between a vertical trunk and lateral shoots are stronger as they approach a 90-degree angle. Narrow branch angles are typically weak and they may break under the weight of developing fruit.

- ☆ Control vegetative growth: Dormant pruning is an invigorating process that stimulates vegetative growth while summer pruning is an invigorating process that reduces growth because it involves removing part of the functional leaf area of the tree.
- ☆ Production: Pruning is done to maintain the right balance between vegetative growth, flower production, and fruiting. Dormant pruning is a reliable crop load management tool.
- ☆ Tree health: Trees are pruned to prevent the spread of disease or insect damage. This includes the removal and destruction of dead and/or infested wood.

Time of Pruning (Fruit Trees)

Dormant Pruning - Winter pruning

- ☆ Generally done after peak cold period is past
- ☆ Plant framework readily visible when defoliated
- ☆ Minimum loss of translocated foods

Summer Pruning

- ☆ Made on new growth
- ☆ Can avoid structural faults before growth is wasted
- ☆ Removes diseased branches in time
- ☆ Better illumination of leaves and fruits
- ☆ Improved assimilation in center of trees
- ☆ Better fruit color, fruit quality, and storage.
- ☆ Slower total growth of trees
- ☆ Less wood production
- ☆ Less total pruning effort and improved work efficiency
- ☆ Possibility of reducing the crown volume of the tree, better utilization of space, more tree per acre
- ☆ Less danger of winter injury

2.8.2 Practice Training

- ☆ Plant management or husbandry is the care and protection of plants so they grow successfully.
- ☆ Plant training is one management practice carried out to control the shape, size and direction of plant growth.
- ☆ It develops desired shape of tree with particular objectives by controlling habit of growth.

- Training is start from nursery stage of plant.
- Some fruit crops like grape vines, ber, fig, guava *etc.* require training.

Features

- Training is originated by the observation that the branches horizontally disposed bear more fruit than the vertical one which are sky to bearing likewise the upper branches bear more fruit then the lower ones.
- Training is mainly concerned with giving a frame and shape to the plant.
- Training determines the general character and even details of plant out line its branching and frame work.
- By training we can keep the plant or vine in a manageable shape and can dispose the branches in desirable direction and position.

Types of Training

- Ground
- Stake and Weave
- Single or double plant staking
- Cages
- Kniffin system
- Umbrella System
- Tree Trellis System
- Over Head "Vertical Line"
- "T" trellis

Ground

Stake and Weave

Single or Double Plant Staking

Cages

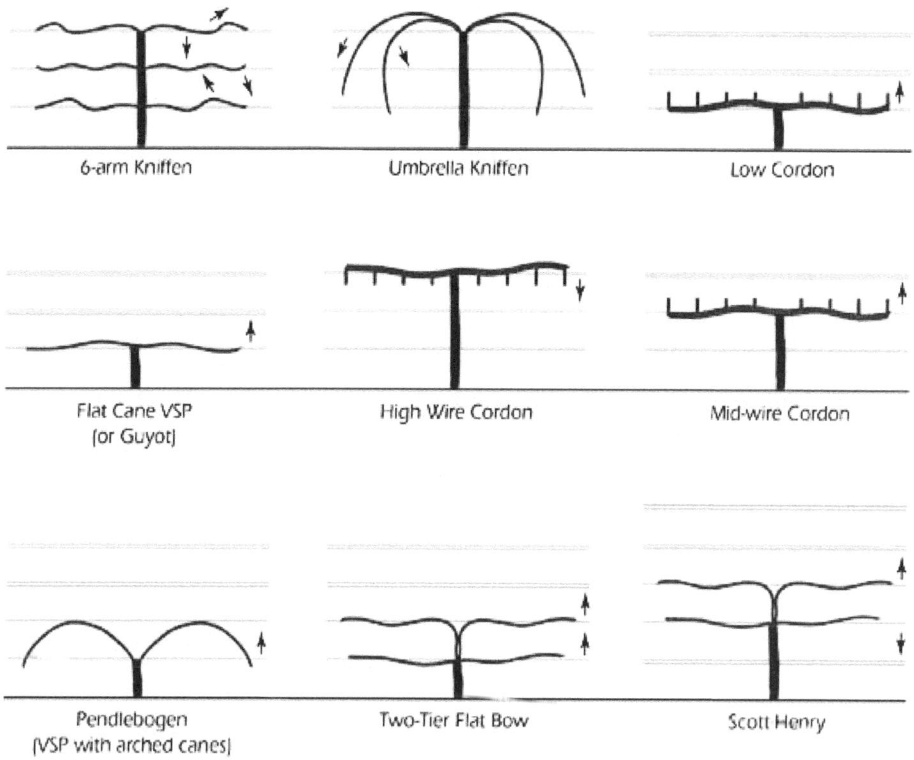

Kniffin System

- ☆ Bower system
- ☆ Inverted V trellis
- ☆ "I" trellis
- ☆ "V" trellis
- ☆ Netting
- ☆ Shift trellis and rotating cross arm trellises

Objects of Training

- ☆ To admit more light and air to the centre of the tree to expose maximum leaf surface to the sun
- ☆ To direct the growth of the tree so that various cultural operations such as spraying, ploughing, harvesting can be performed easily and at lower cost.

- To protect the tree from sun burn and wind damage.
- To secure a balanced distribution of fruit bearing parts of the tree.

Need of Training Plants

You train plants to:

- Improve fruit size and quality
- Develop a framework so the plant gets the maximum amount of light falling on the leaves
- Improve how an area or plant looks
- Make plants easier to manage and maintain the size and shape of the plant
- Improve flower appearance and size
- Protect plants from damage.

Method of Training Plants

Common methods used to train plants include:

- Support including staking
- Stopping
- Disbudding
- Thinning
- Pruning

Plant Support

There are many different systems used to support and grow plants. This is especially true for commercial fruit growing.

Plants are supported to:

- Protect them from the weather, for example, the wind
- Help them grow into the space where they're needed
- Hold leaves up to the light to increase the amount of photosynthesis taking place
- Give easy access of bees and other insects to pollinate the flowers
- To encourage even ripening of fruit for example tomatoes
- To improve the shape so the plants are either easier to manage or look better
- Prevent flowers or fruit from dragging on the ground.

System of Training

Central Leader System

- ☆ In this system the central leader branches are allowed to grow indefinitely so that it will grow more rapidly and vigorously than the side branches and tree became tall.
- ☆ Such a tree bears fruit more near the top.
- ☆ The lower branches are less vigorous and less fruitful.

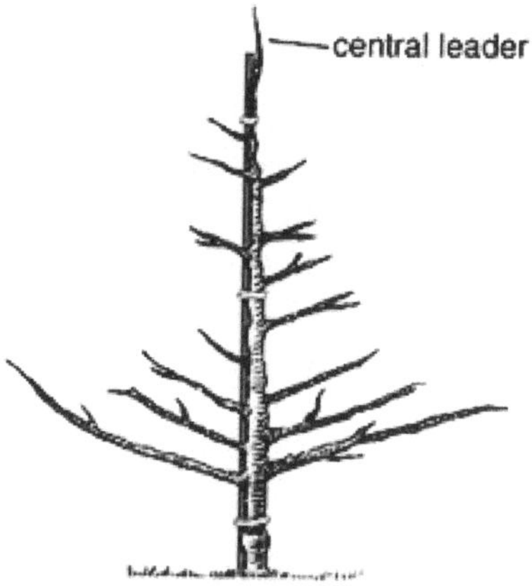

Open Center or Vase System

- ☆ The main stem is allowed to grow only up to a certain height about 1.5 to 1.8 m and then it cut for development of lateral branches.
- ☆ It allows full sunshine to reach each branch.

Delayed Open Center or Modified Leader System

- ☆ It is intermediate between the above systems.
- ☆ It is developed by first training the tree to the leader type by allowing the central axial to grow un-pruned for the first four or five years.
- ☆ The central stem is headed back and lateral branches are allowed to grow as in the open center system.

2.9 Irrigation Management

2.9.1 Water Requirement

Crop water requirements are the depth of water needed to meet the water loss through evapotranspiration of disease free crop, which are grow in large fields under non restricting soil conditions including soil water and fertility and achieving full production potential under the given growing environment.

The crop water need always refers to crop grown under optimal conditions, *i.e.* a uniform crop, actively growing, completely shading the ground, free of diseases, and favorable soil conditions. The crop thus reaches its full production potential under the given environment.

Factors

The crop water need mainly depends on:

Climate

In sunny and hot climate, crops need more water per day than in cloudy and cool climate. The major climatic factors which influence the crop water needs are:

- ☆ Sunshine
- ☆ Temperature
- ☆ Humidity
- ☆ Wind speed

The highest crop water needs are found in areas which are hot, dry, windy and sunny, while the lowest water need are found when it is cool, humid and cloudy with little or no wind. The influence of climate on crop water needs is given by reference crop evapotranspiration which is expressed in millimeters per unit of time, *e.g.* mm/day, mm/month, or mm/season.

There are several methods to determine the reference crop evapotranspiration which are:

- ☆ Experiment using an evaporation pan
- ☆ Theoretical using measured climatic data

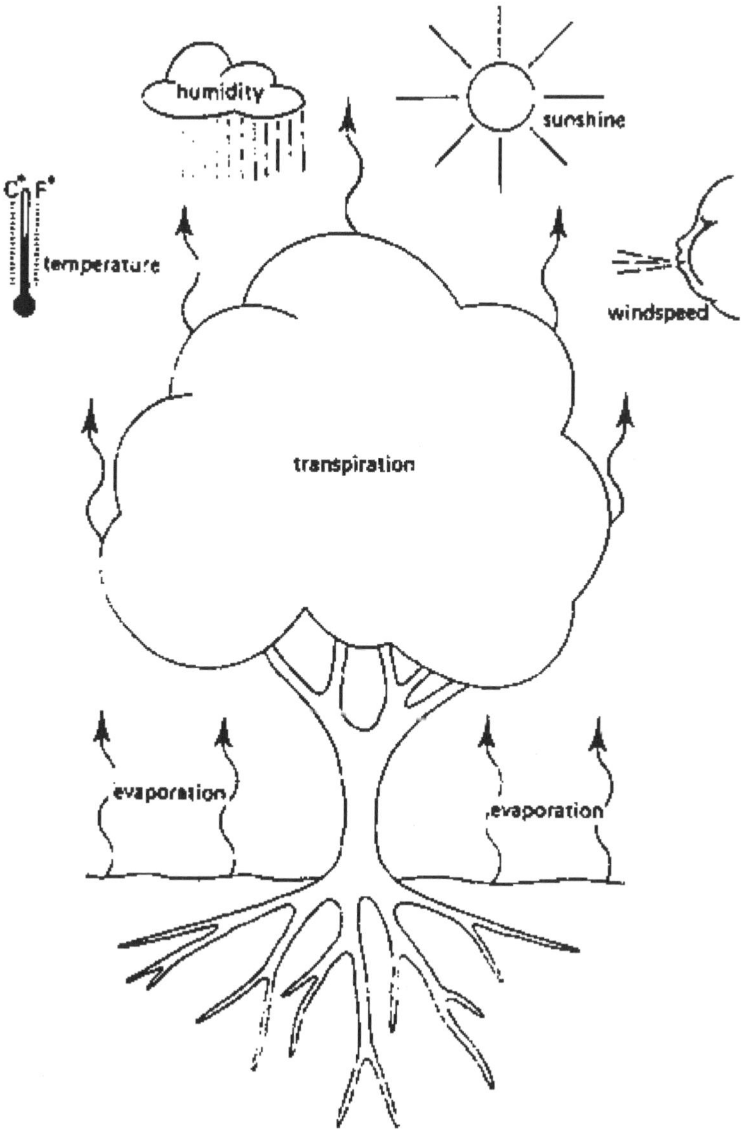

2.9.2 Irrigation Scheduling

There are many factors which a gardener has to consider when developing an irrigation schedule for flower crops which should cover:

- ☆ Water requirements of plants in each zone
- ☆ Local climate and weather
- ☆ Soil composition
- ☆ Sun exposure

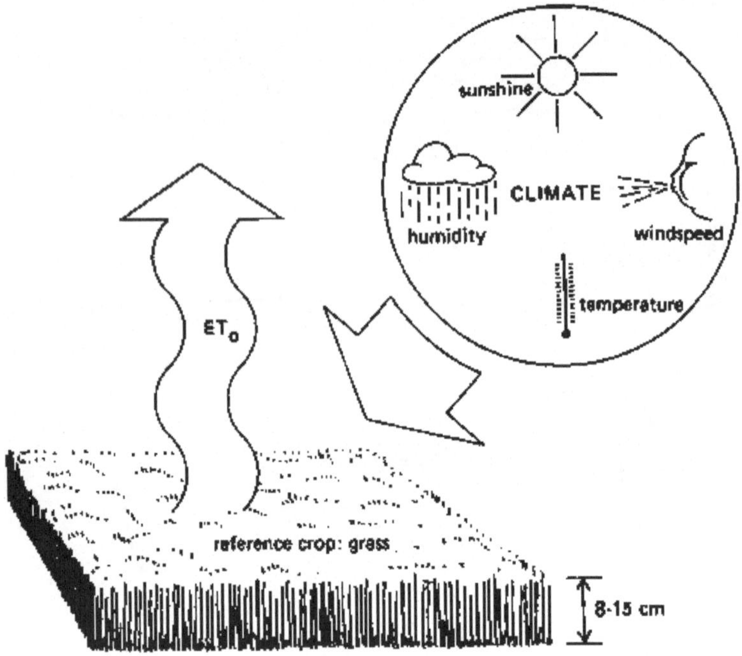

- Slope
- Depth of root zone

Plant Water Requirement

The plant type and size impact how much water a plant will require to remain healthy in the landscape. Information regarding plant water requirements for common landscape plants is often provided by local utilities or extension offices, as well as online. The age of the landscape also impacts the schedule. Immediately following plant installation, landscapes at their early stages require frequent irrigation intervals to establish the plants. Once the landscape is established, however, the schedule should be adjusted to reduce the irrigation frequency appropriate for those plants.

Weather Conditions

Local climate and variations in evapotranspiration play a significant role in determining how irrigation schedules should shift over the season. In addition, specific weather events, such as rainfall, should be considered. If scheduling manually, take these parameters into consideration and create monthly schedules.

A variety of irrigation controllers and sensors on the market can use local conditions to determine watering times and amounts. Water Sense labeled weather-based irrigation controllers have been independently certified for efficiency and performance; they use local weather and landscape conditions to tailor watering schedules to actual conditions on the site. Similarly, soil moisture-based control technologies use moisture sensed in the soil to adapt scheduling to plant water needs. Additional sensors, such as rain, wind, and freeze sensors, can also help delay or prevent irrigation if conditions are not ideal or necessary for watering.

Soil Types

Soil type should also be considered when developing any irrigation schedule. The infiltration rate of the water depends on the composition of the soil. Coarse soil requires frequent irrigation intervals; medium loamy soil requires longer, less frequent watering; and fine clay soils have a high water-holding capacity that requires the least frequent irrigation. If soil type is not taken into account when developing a schedule, water could be wasted due to runoff or deep percolation. For example, if a coarse soil is not irrigated properly, the water will infiltrate to the deep soil quickly and without enough time for the plants to access it. Irrigation water on a clay soil does not infiltrate quickly, and if the system is run too long, the landscape will begin to flood and runoff could occur. Figure 4 demonstrates how water infiltrates into the various soil types over the same irrigation period.

One way to make irrigation in different soil types more efficient is by including cycle and soak in the irrigation schedule. Rather than irrigating each zone fully in a single cycle, the cycle and soak method irrigates each zone in short intervals and allows water to infiltrate between intervals. This is popular for sprinkler irrigation, but can still be useful in micro-irrigation, depending on the flow rate of the system. If the flow rate is slow enough, there will be no pooling on the surface during the entire cycle and no need to adjust the schedule.

Slope and Sunlight

The slope of the landscape and its orientation to the sun should also be considered when developing an irrigation schedule. Care should be taken so as to not schedule long irrigation events on slopes, as this could result in water running off the landscape. Similarly, ponding is more likely to occur at the lower end of the landscape due to runoff from the elevated areas, so lower-flow-rate emitters should be installed, and/or more frequent, shorter events should be scheduled. Regarding sun exposure, plants exposed to intense sunlight, or on slopes with southern or western exposure, might require more irrigation due to the increased evapotranspiration of those plants.

Depth

Lastly, the depth of the root zone should be considered when scheduling irrigation. If irrigation does not reach the full depth of the root system, plants are not receiving enough water to maximize growth. Both plant type and age should be considered when determining root depth and associated schedule parameters.

2.9.3 Sources of Irrigation Water

There are two sources of water

- Canal or surface water
- Ground water-dug wells and bore wells

Canal Water

- The canal water is available from rivers after the construction of dams and stored in dams.
- Rains in seasons cause availability of water and when there are no rains, the problem of water shortage or non-availability of water arises for irrigation as well as for drinking.
- The artificial storages are created by constructing dams across rivers.
- The rivers are small to big, so dams of various sizes and capacity are built which are classified into three categories
    1. Minor-command less than 5000 ha
    2. Medium-Command 5000 to 10,000ha
    3. Major-Command more than 10,000 ha
- The command of minor dams is small, their benefits are accrued in a short period.
- The designing, planning, and construction of large dams take a long time and their benefits are available after a long period.
- Water from the dams is made available for drinking and other human purposes, for industries and for irrigation.
- Huge capital operating and administrative costs are involved in making dam water available for the above-mentioned purposes.

Well Water

- Well, it is the traditional source of water.
- They are constructed by individual farmers from their own funds or bank loans as it does not require much technical know-how.
- Wells can be completed in a short period and their benefits are available immediately.
- The farmers have full control over the use of waste for crops which helps them in deciding what crops to grow, when to irrigate and how much water to apply.
- There is generally no misuse or excessive use of well water and hence no spoilage of lands.

- The command of each well is very small (5 to 10 ha) but the number of wells is very large, so the total area irrigated by wells is substantial.
- As farmers have control on the use of water, delicate crops or crops requiring frequent watering but less water at a time, are grown on well water.
- Fruit, vegetables, and differences in cropping patterns under canal and flower crops are commonly grown on well water.
- There are differences in cropping patterns under canal and well irrigations.

2.10 Pest and Disease Management

2.10.1 Introduction

Pests and diseases often get into healthy gardens/nursery. It is the duty of gardeners to regularly check the nursery plants for spreading diseases or signs of pest damage.

When a flowering plant cannot function normally, it is diseased. The main causes of disease in flower plants are pathogens and environmental factors. Plants have many disease pathogens: viruses, bacteria, fungi, nematodes, mycoplasma similar to organisms, and parasitic higher plants. Fungal pathogens are most prevalent which cause seed rots, seedling damping-off, root rots, foliage diseases, cankers, vascular wilts, diebacks, galls and tumors, trunk rots, and decays of aging trees. Unfavorable weather and environmental factors like temperature and moisture extremes, high winds, or ice can damage plants directly and predispose the plants to pest attack.

2.10.2 Pests Management

A major injury to flower crop is caused by various groups of insects pests such as:

- Major pests
- Minor pests
- Non-insect pests

Army Worm

Symptom of Damage

- Larvae cut the seedlings in large scale
- Severe infestation - cattle grazing appearance to the field.
- They feed gregariously and march from field to field.

Nature of Damage

- Caterpillars march in large numbers in the evening hours and feed on the leaves of paddy seedlings till the morning and hide during daytime.
- They feed gregariously and after feeding the plants in one field march onto the next field. Under severe infestation crop gives the grazed appearance.

Armyworm

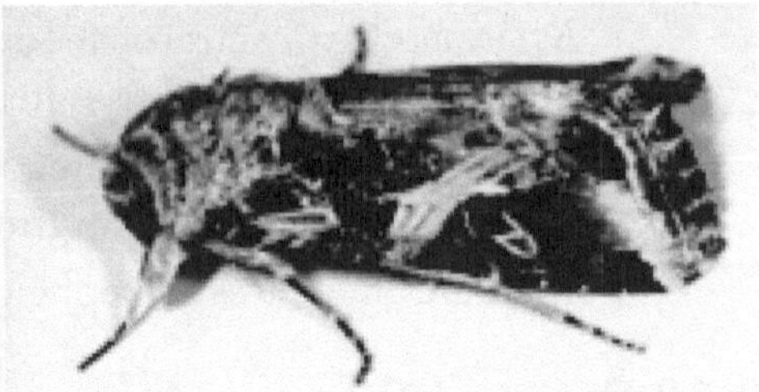

Armyworm Adult

- Attacked plants are reduced to stumps. Fields situated in ill-drained marshy areas attacked earlier than dry ground.
- Damage is severe during July to September.

Identification of Insect Pest

- **Egg:** Eggs are spherical and creamy in colour, which are laid in a group covered over with grey hairs.
- **Larva:** The caterpillars are light green with yellowish white lateral and dorsal stripes in the early stage and later become dark brown or grayish green in colour with a crescent (Semi-circular) shaped black spot on the side of each segment.
- **Pupa:** They pupate inside the soil in earthen cocoons. Pupa is dark brown and measures 16 -17mm long.

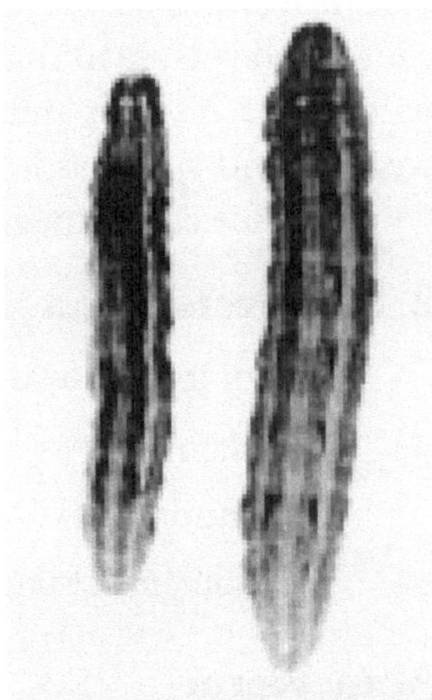

Armyworm larva

- **Adult:** The adult moth is medium sized, stout built dark brown with a conspicuous triangular black spot on the forewings. Hind wings are brownish white with thin black margins.

Management Strategies

Cultural Control

- Flood the area to expose hiding larvae to the surface so that they are picked up by birds.
- Apply kerosene to water to suffocate and kill the larvae.
- Allow birds in field to feed the larvae.

Allow ducks in nursery

Chemical Control

Drain the water from the flower crop and spray chlorpyriphos 20 EC 80 ml

Thrips (*Stenchaetothrips biformis*)

Symptom of Damage

- The flower crop is highly susceptible in nursery stage for first 23-25 days after transplanting/sowing.
- The tips of leaves get rolled longitudinally into needle like outgrowths and turn whitish.
- In severe cases, the lower leaves also turn yellowish.
- The infestation may be rated as mild, if there is less than three needle like leaves and as severe, if there are more than three outgrowths with lower leaves also showing chlorosis and scorching.
- The infestation invariably disappears after sharp showers.

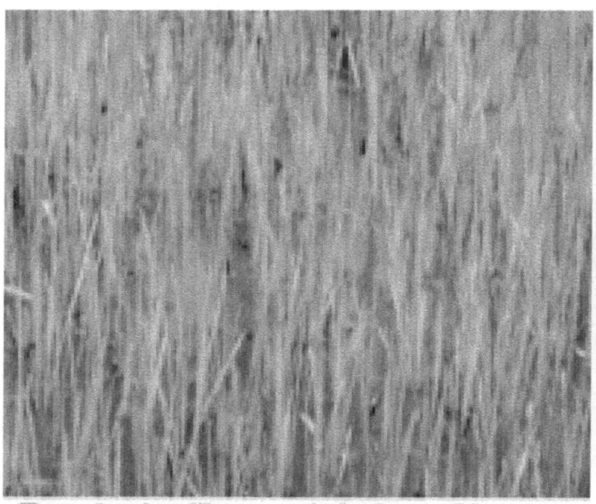

Terminal rolling and drying of leaves

Laceration of tender leaves

Nature of Damage

Both the adults and nymphs lacerate the tender leaves and suck up the plant sap. As a result fine yellowish lines or silvery streaks are seen on the leaves. Later, the leaves curl longitudinally and begin to dry from the tip downwards. In severe cases, the entire nursery may dry up and fail to produce seedling. Sometimes transplanted crop is also affected in the early stages.

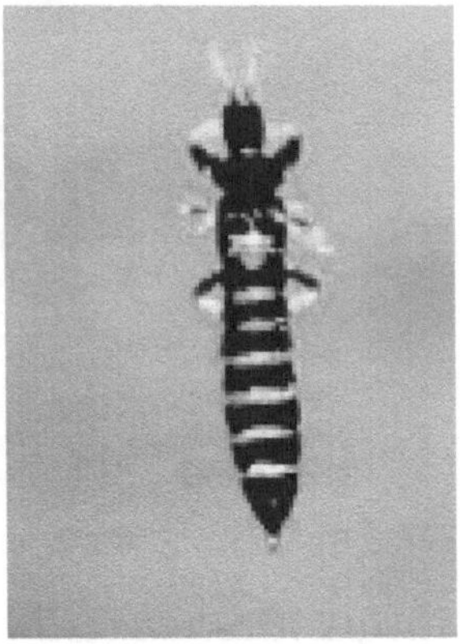

Thrips Adult

Life Stages

- **Egg:** Eggs laid singly in the tissues of the tender leaves on the sides facing the stem. Eggs are hyaline and turn pale yellow as they mature.
- **Nymph:** Newly hatched nymphs are transparent but turn yellowish white after the first moult and possess darker legs, head and antennae.
- **Pupa:** Pupation takes place inside the rolled leaves and appendages and wings are clearly visible.
- **Adult**: Adult is 1mm long, dark brown to black in colour with fringed wings. Male is smaller, more slender than female. It reproduces partheno-genetically since males are seldom seen in the population.

Management Strategies

Cultural Control

- Submerge infested crops intermittently for 1-2 days.
- Drag a wet cloth on the seedlings.

Chemical Control

- Phosphamidon 40SL 50ml (or) Monocrotophos 36SL 40ml

Coccinellid beetle

Predatory thrips

Biological Control

Predatory thrips, coccinellid beetles, anthocorid bugs, and staphylinid beetles are biological control agents that feed on both the larvae and adults.

Green Leafhopper

Symptom of Damage

- ☆ Yellowing of leaves from tip to downwards.
- ☆ Vector for the diseases *viz.*, **Rice tungro virus, rice yellow and transitory yellowing.**

Yellowing of leaves from tip

Green Leaf Hopper nymph

Nature of Damage

Both nymphs and adults suck the plant sap from the leaf and leaf sheath.

Mild infestation reduces the vigor of the plant and the number of reproductive tillers. Heavy infestation causes withering and complete drying of the crop. Plants are predisposed to fungal and bacterial infection through feeding and ovipositional punctures. It also transmits plant diseases such as dwarf, transitory yellowing, yellow dwarf and rice tungro virus (Tungro is transmitted during short feeding period).

Life Stages

- ☆ **Egg:** Greenish transparent eggs are deposited in the midrib of leaf blade or sheath of rice or green grass. They are laid in batches of 10 to 15 arranged in a single row.
- ☆ **Nymph:** The nymphs are soft bodied, yellow white in color. Gradually the color changes to green with five nymphal instars and become adults in 18-20 days.
- ☆ **Adult:** Adults are 3-5 mm long, bright green with variable black markings, wedge shaped with a characteristic diagonal movement. Male insect has a

black spot in middle of the forewings that is absent in females. The insect is active during July to September.

Green Leaf Hopper Adult

- ☆ **Longevity:** Females live for 50-55 days.

Management Strategies

Cultural Control

- ☆ Use resistant varieties like IR 50, CR 1009, Co 46, PTB 2 and PTB 18. Nursery should not be raised near the lamp posts.
- ☆ Apply neem cake @ 12.5 Kg/20 cent nursery as basal dose.

Apply Neem cake

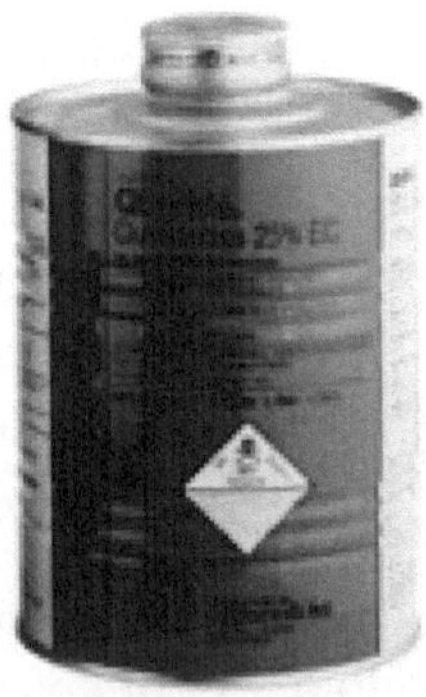

Spray Quinalphos

Chemical Control

- ☆ Spray anyone of the following herbicide: Carbofuran 3 G 3.5 kg (or) Phorate 10 G 1.0 kg (or) Quinalphos 5 G 2.0 kg

Rice Caseworm (Case Worm, *Nymphula depunctalis* (Guen.), Pyraustidae, Lepidoptera)

Symptom of Damage

- ☆ Caterpillars feed on green tissues of the leaves and leave become whitish papery.
- ☆ Tubular cases around the tillers by cutting the apical portion of leaves.
- ☆ Floating of tubular cases on the water.

Whitish papery symptom

Floating of tubular cases

Nature of Damage

The caterpillar cuts a piece of leaf roll it longitudinally into a tubular structure and remains inside. It feeds by scraping the green tissue of the leaf. The cases often float in the water.

Identification of Insect Pest

- ☆ **Egg:** Eggs are light yellow, disc like, smooth and irregular in shape. They are laid on the underside of the leaves floating on the water.

Caseworm Adult

- **Larva:** Pale translucent green with orange head. It has filamentous gills on the sides of the body that helps to lead a semi aquatic life
- **Pupa:** It pupates inside the leaf case. Fresh pupae are milky white, which gradually turn to light yellow.
- **Adult**: Moth is delicate white moth with pale brown wavy markings. The adult is a small delicate moth having white wings speckled with pale brown wavy markings. Females are larger than males. Egg laying takes place in the night.

Management Strategies

Cultural Control

- Mix 250 ml of kerosene to the standing water
- Dislodge the cases by passing a rope and drain water
- Collect the cases and destroy

Apply kerosene to water

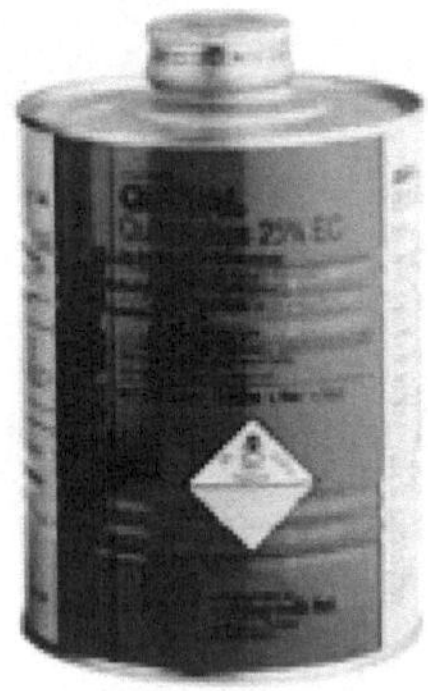
Spray Quinalphos

Chemical Control

- Spray any one of the following herbicide : Monocrotophos 36 SL 40 ml (or) Quinalphos 25 EC 80 ml.

2.10.3 Steps for Pest Management

- Identify bugs.
- Determine the severity of the problem by inspecting undersides of leaves and new leaves and check for disease spreading.
- Make a plan.
- Consider what controls to choose (organic pesticide at the last).
- Keeping notes on spray and pest problems.

- Consider increasing the diversity of flowers in the garden which will attract beneficial insects.

2.10.4 Pests and its Controls

- Aphid: Apply a strong spray of water or insecticidal soap and allow beneficial insects such as ladybugs or lacewings.
- Flea Beetle: Use agricultural row cover to protect young crops by sticky traps which will help in monitoring the presence of beetles.
- Cucumber Beetle: Use agricultural row cover when crops sprout and remove the cover when flowers appear so that they can be pollinated.
- Imported Cabbage Worm: Spray Bacillus thurengiensis at first sign of cabbage worms
- Leaf Hopper: Row covers, sticky yellow traps, insecticidal soap.
- Stink Bugs and Harlequin Beetles: Hand pick, especially eggs.

2.10.5 Disease Management

Objectives

The main objective of disease management is to prevent or minimize losses while preserving plant quality. Absolute disease control is rarely achieved or even attempted. More often, management efforts are directed towards preventing disease or reducing it to the status of a tolerable nuisance. In most instances, forest disease management requires preventive methods over a long period of time and considers the stand as a whole rather than specific diseased individuals. Management measures must be economically feasible - expenditures must not exceed the expected benefits. Direct control of disease in the forest is limited by many factors, including:

- The vast areas involved
- The inaccessibility of many stands
- The long life cycle of trees
- The relatively low per acre or per individual tree values
- Spraying, dusting or other direct control procedures commonly employed with high-value of nursery crops and valuable seed orchards are rarely applicable.

Principles

- Avoidance: Use or handling of environmental factors in the implementation of plant protection in order to profit from the absence, inability to infect or the interruption of the biological cycle of the pathogen.
- Exclusion: attempts to prevent the introduction or establishment of a pathogen in an uncontaminated area.

- Eradication: elimination of a pathogen from an area in which it is already established.
- Protection: obstruction of economic damage in a crop by a pathogen by imposing a chemical or physical barrier between the inoculum and the plant. Protective measures should be taken either just before the appearance of the inoculum in plants or at the beginning of the incubation period.
- Therapy: application of physical or chemical means so as to destroy the pathogen into the plant.
- Development of resistance: handling of the morphology or physiology of a crop by improving the methods of selection or hybridization so that the pathogen cannot be established.

Chapter 3

Harvest and Postharvest Management in Floriculture

3.1 Practicing Flower Harvesting

3.1.1 Introduction

Harvesting of flowers on a large scale is done by using tools that make the harvesting process more efficient and less time-consuming. Also, things that grower/floriculturist must be aware of when harvesting flowers are designed for flower markets is the possibility of damaging the soft, delicate stems during the harvesting thereof.

To reduce such damage of flower crop stems, it is recommended that, grower/floriculturists must use a pruning tool that incorporates cut-and-hold devices on the cutting head. Such a device allows the floriculturist to hold onto the stem of the flower once it is cut, using only the hand that is holding the tool. Further, the stem can be placed either in picking buckets with water or in the user's arms, all in one motion.

3.1.2 Harvesting Time

The best time to harvest the flower crops is late morning, just after the evaporation of dew from the leaves. Many flowers do best when cut slightly immature, with the bud not fully open, as the flower will continue to open once cut. Often, fully open flowers will drop their petals as they dry.

The timing of harvesting flowers depends on:

- ☆ Point in flower's maturation process
- ☆ Time of day to harvest

The best time of harvest is the coolest part of the day and when there is no surface water from dew or rain on the plants which can be in cool of the morning or late afternoon or evening as the plants have stored carbohydrates from the day which will provide a food reserve for the plant material. Harvesting too late will lower the vase life of flower while harvesting too early will lead to the immaturity of flower that hasn't had sufficient time to develop, and may never properly bloom.

Features of Time of Harvesting

- The time of crop harvest is an important strategy that influences the flower crop yield, quality, and storage of the crop.
- Premature or delayed harvest flower crops will adversely affect the quality of flower production.
- Pre-mature or immature flower harvest is desirable for certain products that are preferred juicy and succulent.
- Produce may become more fibrous and tough or hard if harvested at full maturity.
- If crops are left in the field for too long after maturity, the flower crop will be predisposed to pests and germination is inevitable due to vagaries of weather.
- There are different stages for various flower crops, which demarked as maturity stages, harvesting in those stages gives high income and good quality produce.
- Gardeners/Farmers are following traditional techniques for harvesting the flower crops produce.
- Flower crops are harvested when they are said to be mature.
- The ideal time to harvest flower crops depends on factors such as economic part, utilization of product and postharvest storage.

Economic Part

- The economic product could be grain, root, leaf, stem or other parts for flower crops.
- Translocation of stored food from parts of economic importance to other parts will reduce the yield of desired products.
- Flower crops to be harvested if the desired product is at its maximum quality and quantity.

Utilization

- Economic product results same but are desirable to harvest in fresh or dry.
- Flower crops can be harvested fresh or dry as per the intended use.

- The purpose of growing flower crops determines when it is best to harvest for having max quality.

Physiological Maturity

- Flower crop is at physiological maturity if the translocation of photosynthates is stopped to the economic part.
- It shows the developmental stage after which no further increase in dry matter in the economic part.
- In flowers, the moisture content of grains is very high during the milking stage and decreases due to the accumulation of photosynthates.

3.2 Identify Harvesting Stage

3.2.1 Introduction

- Stage and time of harvest of flower crops play an important role in attaining good crop produce.
- Premature or delayed crop harvest will adversely affect the quality and quantity of flower production.
- Factors considered for the time of harvest include economic part aimed, product utility and postharvest storage.
- All flower crops are having certain criteria for harvesting the flower produce.
- There are different methods of harvesting of flower crops such as manual and mechanical methods which are practiced as per the purpose, time and labor.

3.2.2 Stages

When the flower plants switch from vegetation to the flowering stage, the growth will get slower. If the flowering plant starts to develop flowers, it is treated differently as a change in light and nutrients for the flowering stage. If flower plants have been growing and prospering in the vegging stage, then such plants will be prone to problems, as all happens from here on out will affect flower crop yield for better or worse.

The complete flowering stage will be of eight weeks which comprises of:

- transition
- first buds
- growth of buds
- buds ripen
- harvest

Stage 1: Entering Flowering Stage

- To introduce plants to this stage in the life cycle, floriculturist/grower has to make sure that they use the right flowering growing lights.
- Gradually changing light cycles till grower reaches 12/12 hours ratio of light and darkness, the growing indoor flower gives a good opportunity to create an environment that suits plants the best.
- It is important to take advantage of indoor growing plants instead of being dependent on the time of sun, better the flowering stages, better will be the yield.
- It is seen that the initial 1-3 weeks of the flowering stage are a transition phase where plants will grow.
- In this, plants could double their height at a certain point as per the type of strain to be used.

Stage 2: Spotting First Flower

- Before spotting flowers, floriculturists/growers still want to support plants in the growth stage as much as possible.
- Once flower shows, grower/floriculturists can optimize for light and nutrients and should switch to nutrients for the flowering stage.
- In this, only female plants produce buds, as a grower should make sure to have female plants using feminized seeds.

- ☆ In 3-4 weeks, initial real buds show instead of hairy pistils, where the growth of flower plants slows down a lot.

Stage 3: Bud Pistillation

- ☆ In this, white pistils stick in all directions where buds get bigger day by day.

3.3 Handling Harvested Flowers

3.3.1 Harvesting of Flowers

Harvesting is normally done by hand using shears or a sharp knife. Simple mechanical aids are used to harvest some crops, like hook-shaped "comma" which allows harvest without stooping, and rose shears which grip the flower stem after it has been cut, allowing it to withdraw single-handedly from the bench. At no time should harvested flowers be placed on the ground, due to the danger of contaminating flowers with disease organisms.

It is observed that flowers remain alive even after harvest and continue their metabolic activities which result in:

- ☆ Depletion of carbohydrates
- ☆ The rise in temperature and respiration rates
- ☆ Rapid deterioration due to micro-organisms
- ☆ Water stress
- ☆ Increased accumulation of ethylene

3.3.2 Flower Care during Harvest

The quality of flowers needs to be best during the harvest. After harvest, flower quality can only decline as the flower matures and the stem dies. Great care and attention to the smallest detail must be taken by the gardener to slow the decline of the flowering stem during harvesting. The rate at which flowering stem declines is dependent on tissue temperature and water status of flower, stem and foliage. High temperature increases the rate of decline. It is not practical to pull a refrigerated cooler through the fields while harvesting, but other practices can be of value. Plant tissue temperatures will be lower during cooler parts of the day. Harvesting early, during the cool of the morning, will help in maintaining quality. Do not harvest until after any dew has evaporated off the flowers. Damp flowers packaged or placed into refrigerated storage may develop Botrytis, a fungal disease which attacks and destroys salability of the flowers.

To maintain cooler tissue temperatures, harvested flowering stems should be kept in a shaded area until transported to grading/packing area and refrigerated storage. For this, the covered harvest trailer is a good method of providing such shade.

Also, to maintain the quality of harvested flowering stem to maintain the water status of plant tissue, the harvested stems should be left to dry till brought into grading/packing shed and get wilt to a point beyond an ability to recover. Cut stems should be placed in either water or fresh flower food. The typical flower food solution contains water; simple sugar that serves as a food source; chemical to prevent or retard the growth of microorganisms which can plug the conductive tissue; and an ingredient to acidify the water, typically citric acid. Fresh flower preservatives are available from several companies.

3.3.3 Keeping Harvested Flowers

Plastic containers are best for use to hold harvested flowers while metal containers need to be avoided due to the possibility of rust that forms and block the stems' conductive tissue. Chemicals in floral preservatives react with metal containers.

Harvested containers need to be stored in an area and manner which will keep them clean when they are not used. Floriculturists can stack the plastic buckets by inserting them one on another, so that the maximum number of buckets can be stored in the smallest possible space. Such a container storage system has many serious deficiencies, as the dirtiest part of many containers is the bottom, which is an outside surface that sits on the floor or ground. By keeping containers into each other, will make the dirty surface placed in another container, which creates the potential for dirt to be inside each and every container.

3.4 Grading of different Flower

3.4.1 Quality Grades of Flower

Standard flower grades will help the growers/floriculturist to improve the marketability of their flower. Such grading standards help to remove the human elements in grading by using inflexible standards. The idea of grading will make the product easier to sell at all levels.

The flower quality depends on the systems of:

- Pre-harvest
- Harvest
- Postharvest

3.4.2 Factors on which Flower Grading Depends

The flowers quality standards will depend on:

- Shape and size of flower bud
- Turgidity or freshness
- Vase life or keeping quality
- Stem length
- Stem strength
- Condition of calyx, color, fragrance
- Development and condition of foliage
- Foliage free from dust and spray residues
- Absence of insect/diseases
- Status of physical defects

It is observed that the highest quality flowers will be rated as select while flower with a minute flaw, is rated as fancy. If petal in flower is less with shorter stems, then such flower will be graded as grades A, B, and C. Flowers which are available in the open market are likely to be with grades A, B, and C.

3.4.3 Types of Grading Systems

There is no definite flower quality standard for grading of different flowers which varies from country to country and grower to grower, so for grower/floriculturist, it is better to keep colors separate in bunches and containers, to identify different flower grades. The flowers along with their attached stems should be free from disease and pests, so every flower/bunch in a pack requires uniform weight, so based on the quality of flower crops, flowers can be graded as:

- Extra class
- First-class
- Second class

3.4.4 Different Flower Grading

- Rose: The rose flowers are graded based on lengths of stems.
- Gerbera: The Gerbera is graded as per the stem length, flower diameter, and maturity.
- Orchids: The grading of orchid is based on the length of the flower spike, the number of opened flowers and buds.
- Chrysanthemums: The chrysanthemums are graded based on stem length, flower appearance, number of flowers, stem straightness, color and freshness of flowers and are graded into Blue, Red, Green, and Yellow, whereas spray types are graded into Gold, Silver, and Bronze based on the quality parameters.
- Gladiolus: The spikes are graded in three lots, as per overall quality, variety, color spike, length and number of florets per spike.
- Tuberose: The tuberose flower spikes are graded as per the stalk length, length of rachis, number of flowers per spike and weight of spike.

3.5 Use of Chemicals for Enhancing Vase Life of Flower

3.5.1 Introduction

- There are many varieties of flowers that have longer vase life while some have shorter vase life. The flowers with a shorter vase life of 4-5 days can be cosmos, sweet peas, zinnias, *etc.* while flowers that last long in vase for around two weeks can be lisianthus, feverfew, strawflower, bells, *etc.*

- ☆ The average vase life of flowers was counted from the day of transferring the spikes in the solutions and was assessed to be terminated when flowers lost their ornamental/display value.
- ☆ The criterion for the end of vase life was the time when flowers show symptoms of petals wilting or curling.
- ☆ Vase life was the period from the time of putting the flowers in preservative solution till the end of vase life.

3.5.2 Factors Determine the Vase Life of Flowers

- ☆ Growing conditions: Light, temperature, relative humidity, fertilization, and crop protection have a marked effect on the quality of flowers once harvest.
- ☆ Postharvest conditions: Temperature and humidity, in particular, determine whether the cut flower will be delivered to the consumer in good condition after the harvest. Higher temperatures promote flower development. Humidity plays a direct role in the condensation process and wide fluctuations in this and temperature promotes the growth of the fungus Botrytis. Using the most suitable postharvest treatment for each flower type is one of the measures needed to get the flowers through the distribution chain intact.
- ☆ Hygiene: Poor hygiene measures promote the development of micro-organisms, which cause the vase water to become cloudy, dirty and ultimately four smells.
- ☆ Distribution/transport time: Flowers come from all corners across the world and are shipped across huge distances by road, air or sea. Climate control, good packaging, appropriate postharvest treatments and speed are important in conserving the quality of flowers and foliage.

3.5.3 Chemical Requirement

- ☆ Adding chemical preservatives to the holding solution will prolong the vase life of flowers.
- ☆ The holding solutions should contain components such as sugar and germicides.
- ☆ The sugar provides a respiratory substrate, while germicides control harmful bacteria and prevent the plugging of conducting tissues.
- ☆ Among all different types of sugars, sucrose results as a commonly used sugar in prolonging the vase life of flowers.
- ☆ The exogenous application of sucrose supplies the flowers with much-needed substrates for respiration which allows the flower to harvest at the bud stage to open, which otherwise could not occur naturally.

- ☆ Silver thiosulfate acts as an ethylene antagonist which reduces ethylene production and respiration that helps in extending flower longevity.
- ☆ Vase life increases three times by solution having 200 ppm silver nitrate and 4 mM silver thiosulfate.
- ☆ With the application of sugars and antimicrobial agents, the vase life of flower increases to 22 per cent and improved in spikes quality.
- ☆ The postharvest life of flowers is dependent on carbohydrate status and the amount of metabolic sugars results as factors that affect the rate of senescence.
- ☆ Keeping the flower in vase solutions having sucrose, extend the vase-life.
- ☆ As seen, the addition of sugar to vase solution improves the intensity of petal color but did not improve bud opening, longevity or size of non-cold-stored.
- ☆ The addition of sugar to vase solution of defoliated stems not only restored the color on petals but increases the size of open flowers.
- ☆ Vase life increased significantly if glucose concentration increased from 10-15 percent.
- ☆ Vase life was significantly affected due to the harvesting of flowers at the tight bud and half-opened bud stage.

3.6 Value Addition of Flowers

3.6.1 Introduction

With high perishable nature, many flowers gets wasted, so value addition is good way to avoid wastage and utilize glut. Value addition shows increase in value of raw flower product anytime between harvesting and sale of final product which can be done by changing processing.

As the flowers are the utmost perishable horticultural farm produce, so there remains few hindrance in proper marketing that follows standard postharvest management practices by common gardeners/floriculturist.

As seen, value addition by agro-industries is other arena for proper use of fresh ornamentals in either garden-fresh or processed form. There are different kinds of value-added products that are nowadays formulated and marketed by companies like:

- ☆ Essential oils,
- ☆ Flavours,
- ☆ Fragrance,
- ☆ Pharmaceutical and nutraceutical compounds
- ☆ Insecticidal and nematicidal compounds
- ☆ Pigments and natural dye
- ☆ Vanilla-based products
- ☆ Gulkand
- ☆ Rose water *etc.*

In flowers, value addition serves as a process where same volume of primary product, high price is realized using processing, packing, upgrading the quality or other such methods.

Value addition includes preparation of products from fresh flowers like cut flower arrangements, bouquets, garlands, corasage, wreaths, buttonholes, floral *etc.*

3.6.2 Objectives of Value Addition

The main objectives of flower value addition is to:

- ☆ Improve the profitability of gardeners/farmers
- ☆ Empowering farmers/gardeners and others by gainful employment opportunities
- ☆ Provide good quality, safe and branded products to consumers
- ☆ Emphasize primary and secondary processing
- ☆ Reduce postharvest losses

3.6.3 Need for Value Addition

- ☆ Lowers import and meeting export demands
- ☆ Increasing foreign exchange
- ☆ Encouraging growth of subsidiary industries
- ☆ Lowering economic risk of marketing
- ☆ Increasing opportunities for small farms/shops by development of markets
- ☆ Diversify economic base of rural communities
- ☆ Increases farmers financial stability

3.6.4 Value Addition Products

- ☆ In addition to number of products such as essential oil, perfumes, insect repellants, cosmetics, pigments like xanthophylls, carotenoids, luteins, anthocyanins, *etc.* that have neutraceutical properties and pharmaceutical compounds can be prepared from the flowers.
- ☆ For successful value-addition, it is recommended that carefully identification of goods uses local resources and fulfill the gap in the market.
- ☆ Adding value adds to the cost of production, but careful planning and test marketing can significantly increase the net return of small-scale floriculture enterprise.
- ☆ Value-addition allows high premium to floriculturist/grower, while providing more acceptable quality products for domestic and export market and allow customers to buy such products.

3.6.5 Cause of Value Addition

There are three main reasons for value addition:

- ☆ Use of excess produce and removal of waste
- ☆ Stabilizing farm income by adding profits
- ☆ Affording shop/outlet for creativity

3.6.6 Modes of Value Addition

There are different modes of value addition such as:

Fresh Flower Products

- It includes two types of flower arrangements; bouquets, baskets, bunch, buttonaire, corsage *etc.* and loose flower products like garland, floral strings, pomanders, wreaths, floral jewelry *etc.*

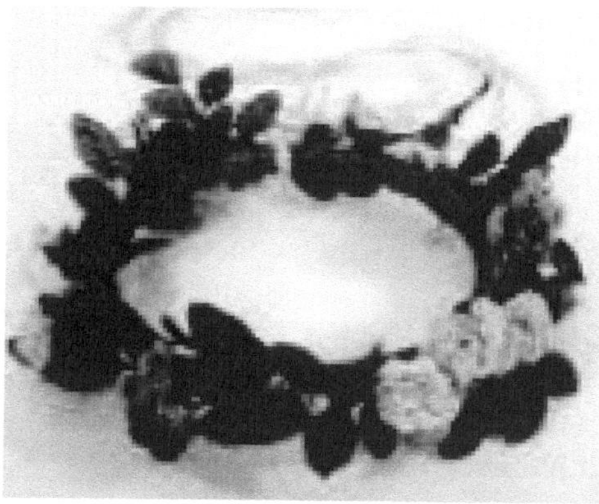

- The floral jewelry concept became popular among brides on mehandi event in marriages, as floral jewelry matches with dress and light that is convenient to use with no chances of jewelry theft.
- These flower have limited shelf life, so it can be used/enjoyed from 2-15 days as per the prevailing climatic conditions and species used.

Dry Flower Products

- Dried flowers are exported from many years to various countries.
- The dry flower export from the country has a contribution of around 70 per cent of total export which makes the great scope of drying the flowers that grow in the local area which convert in everlasting value-added products like dry flower arrangements, baskets, bouquets, table arrangements, *etc.*

- With different dry flower products, pot pouri serves as main dry flower export item along with press dried flower products such as greeting cards, bookmarks, swaths, paperweights, wall hangings, tabletops, table mats, *etc.*
- Many dry flower products have a shelf life of 6-9 months, so the customer can enjoy such flower products for a longer time and can use the maximum benefit of the amount spent.

Processed Products

- It covers essential oils, absolutes, concretes, petal jam, jelly, ready to serve beverages, wine, floral tea, rosehip juice, poultry feed, insect repellent, floral dyes, petal embedded handmade paper, cosmetics, rose water, rose cream, *etc.*

Pharmaceutical and Nutraceutical Compounds

- With globalization, international competition increases and people become conscious, so the use of natural ingredients increases in the food and pharmaceutical industry.
- Flowers pigments such as carotenoid, xanthophyll, anthocyanin, *etc.* are being used in the food industry and compounds such as lutein, vincristine, catharanthine, vitamin C extracted from rose hips are used in the pharmaceutical industry.
- Carotenoids are the main source of pigment for poultry feed which is used in feed for intensification of the yellow color of egg yolk and broiler skin, thereby increases lutein content, which is good for health.
- There results in scope of converting flowers in different value-added products as required, desire and as per the condition of flowers.

3.6.7 Benefits of Flowers Value addition

- ☆ Improving quality
- ☆ Increasing sale
- ☆ Achieving high profit
- ☆ Lowers postharvest loss
- ☆ Gives demand for high quality
- ☆ Use of unsold flowers
- ☆ Increasing in export
- ☆ Generating employment

3.7 Packing of Flowers

3.7.1 Introduction

Good packing is of great importance to floriculturist/flower grower as he/she needs to make sure that the flowers do not arrive crushed or bruised. If flowers are delivered to recipient with good packing, it gives an overall impression to customer about the person or shop.

Using wrong type or size of packing material may result in damage or breakage, which results in customer dissatisfaction and can lead to business loss. Floriculturist, before using packing supplies should need to think about the particular requirements of items involved.

Packing for flowers can be very different, as there results different ways of decorating a composition. It does not matter why flowers are presented as gift, where it is more important to provide a presentable bouquet to impress the recipient as beauty of flower arrangement or bouquet can be judged on the packaging.

3.7.2 Types

There are many different types of packing such as distribution package or transport package, that are used to ship, store, and handle a flower product or inner packages. Packaging may be looked at in relation to the type of product being packaged which can be seeds, flower packaging, over-the-counter packaging for retail.

Packing Containers

- There are different shapes of packing containers for flowers, which are long and flat with square or rectangular design.
- In this, flower heads can be placed at both ends of the container for better use of space and are covered with newspaper to protect flower heads.
- To avoid longitudinal slip, packers uses cleats which is foam or newspaper-covered wood pieces that are placed over the product, pushed down, and stapled in each side of box.
- Padded metal straps, high density polyethylene blocks, and cardboard tubes can also be used as cleats.
- In this, heads of flowers is placed 5 – 10 cm from the end of box to allow effective pre-cooling and to eliminate danger of petal bruising.

- Cushioning materials like shredded paper, paper wool and wood wool may be distributed between packed flowers to further reduce damage.

3.7.3 Function/Requirements

The main objective of flowers packing is to increase the sales and profit of a product. The functions and requirements of packaging are:

- Protecting flowers against physical/chemical damage during storage and distribution.

- Making the product distinguished, competitive and attractive.
- Presenting product in good hygienic and attractive way
- Providing minimum inconvenience and maximum satisfaction to customers/users.
- Saving time, labour, space while handling and storing of flowers
- Help in selling of product
- Economical

3.7.4 Features of Packaging Systems

- Inertness: Compatibility with the product to maintain it in a satisfactory state during storage and transit until it reaches to a retailer
- Fitness for consumption: To provide product security during storage and transit
- Environmental protection: It should able to provide environmental protection against the high risk of adverse interchange with gases
- Mechanical protection: Against handling and transit hazards
- Cost: Economical and available in commercial market
- Decorative: It should be attractive provide adequate surface for identification of the product. It should help in selling the product *i.e.* it should give description, introduction
- Bulk density: It should give idea about weight, volume, and number

3.7.5 Importance of Packaging

The collection of flowers is joyful and by seeing them together in an arranged fashion will give grace. Flower packing will help in:

- Easy Transportation: As flowers are fragile, so having them in package is a safe way to transport them safely and to maintain their beauty.
- Appealing: Packed flowers allow collective arrangement. Seeing a whole collection is splendid as it is delightful.
- Good Presentation: A bouquet is always a delight, so in it, flowers packed are always splendid.

3.8 Tools for Flower Harvesting

Professional florists and floral enthusiasts can learn how to cut/harvest flowers with increase efficiency and more floral display time using certain harvesting tools, which can be:

Wire Cutters

- ☆ It is used to cut chicken wire, which is used in floral designs which supports the stems of flowers

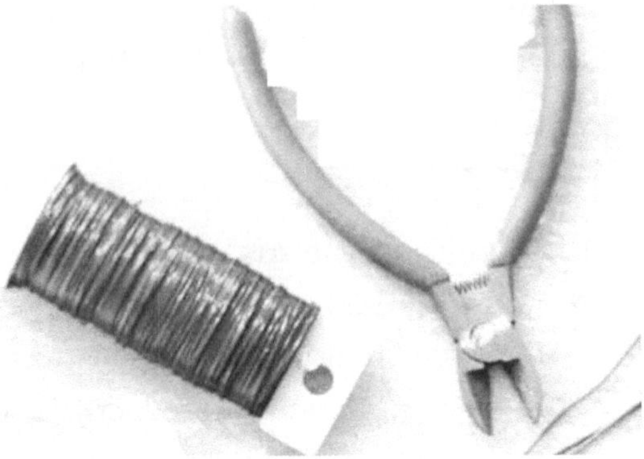

Floral Snips

- ☆ It is lightweight, super sharp and is perfect shaped to rest in palm of the hand.
- ☆ It has thin pointed blades, which is ideal for harvesting flowers like sweet peas, ranunculus, dahlias and zinnias.

Pruning Shears

- ☆ It is an ideal tool for harvesting flower crops with woody stems like peppers, pumpkins and squash.
- ☆ It is a type of scissor used for cutting plant stems.
- ☆ Floral designers use pruners for cutting through tough, woody stems that are not easily cut with knives.
- ☆ It has ultra-sharp blades and sturdy handgrips which makes it ideal for pruning of woody stems like roses, shrubs and trees

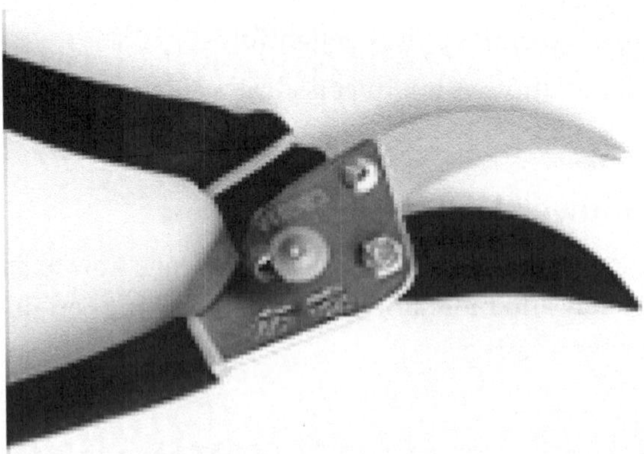

Hasami

- ☆ It is a scissor which is very sharp and made from hardened steel.
- ☆ It has relative short blades and has no spring mechanism to keep it open.

Sharp Knife

- ☆ It is used for cutting vegetables like squash, broccoli and cabbage.
- ☆ It is a sharp pruning knife used to cut flower stem.

Hori Hori Knife

- ☆ It is a tool used by landscapers for cutting woody stems, transplanting, mixing soil amendments *etc.*
- ☆ It is used for digging, sawing and dividing plants with ease.
- ☆ It is the most versatile tool for gardening and is used for transplanting, weeding, opening plastic bags and cutting twine.

Spade

- ☆ It is used for harvesting root veggies.
- ☆ It should be used correctly as it can lead to damage of flower crops.

Loppers

- ☆ It is a heavy duty, long-handled which is an essential tool for flowering branches like crab apples, viburnum and foraged lilacs.

Stem Cutter

- ☆ It is an essential tool for florists and flower farmers alike.
- ☆ It gives bunch of flowers a quick trim before design work, and helps to prolong vase life.
- ☆ Such tool saves time and wrist-ache.

3.9 Storage

3.9.1 Introduction

- Once the flower is harvested, it is then stored for further use.
- Harvested produce often requires some form of storage at site of production, before disposal.
- As seen, respiration in flower continues after it is cut.
- The rate of respiration corresponds to changes in metabolism which leads to senescence.
- The rate of respiration also related to storage temperature, where higher temperature leads to more rate of respiration and subsequent decline in flower quality.
- For most crops, a temperature of 40° to 60°F is desirable for short-term and 32° to 35°F for long-term storage.
- The produce may deteriorate in storage if harvested at improper moisture content.
- There must be supplementary drying facilities for drying the product to safe moisture content for storage.

3.9.2 Air Conditioning Cold Room

Refrigerated storage or cold room can be used for storing flowers. In this, a cabinet is good for small growers as part of a variety of products while for big retail flower growers and whole sellers, walk-in coolers can be fabricated on site. In determining size, consideration should be given to provide enough storage capacity for 24 to 48 hours for retail operations and for longer periods for wholesale operations.

The air conditioning cold room is used for long-term storage of fresh fruits, vegetables, flowers, *etc.* To control temperature and moisture in cold room, respiration of plants is important which regulate and control O2, CO, N2, and ethylene, and inhibits respiration and metabolism of other plants in a dormant state for achieving long-term storage.

Advantages

- It stores fruits, vegetables, flowers, *etc.* for a longer period.
- It gives a good preservation effect.
- It keeps fruits, vegetables, flowers, *etc.* on storage rack for a long time.
- It is a green environmental way of storage.
- It has a low consumption.

3.10 Transportation

3.9.1 Temperature Management before Loading

Ideally, flowers would be packed and cooled at the production location and loaded into LD-3 containers or made into pallets on LD-9 sheets before being transported in a refrigerated vehicle to the freight forwarder or air cargo facility.

Whenever active refrigeration is not available, flowers should be kept cold by the following measures:

- ✰ Consolidation into pallets to reduce the surface/volume ratio
- ✰ Reduction in radiative heat gain by holding in a shaded location
- ✰ Reduction of air infiltration by closing precooling vents and using pallet covers.
- ✰ Pallet covers should preferably be white or foil-backed and have the additional benefit of reducing water loss during air transport.

Flowers cooled to 0°C and packed in a large pallet will gain about 1°C per hour due to the heat generated by the process of respiration. As the temperature rises, respiration rises, and the heat gain increases too so that by the time the flowers are at 10°C they may gain 3°C per hour, and once they are at 20°C, the heat gain can be as much as 10°C per hour. Flower boxes unloaded from aircraft pallets have been probed at temperatures as high as 55°C, indicating the runaway danger of inadequate temperature management and delays in transport.

3.10.2 Transportation System

- ✰ Flowering stems to be sold as a fresh product must be transported from the plant to the harvester's arms, to aisle between production beds, to the roadway, to holding/storage facility, to a grading/packing area, to refrigerated storage and finally to delivery truck.
- ✰ Stems which will be marketed as preserved materials must also be transported in and out of processing areas, so for every product movement activities, a system needs to be developed.
- ✰ A transport system and all subsystems must be developed specifically for each company.
- ✰ The site plan, topography, production system, existing equipment and facilities, workforce and the intended market all affect the transport labor system design.
- ✰ Transport systems must be integrated with all other aspects of the company.
- ✰ The linkages between each activity are critical.
- ✰ Design each subsystem so as to enhance the performance or lower the cost of the next activity in the system.

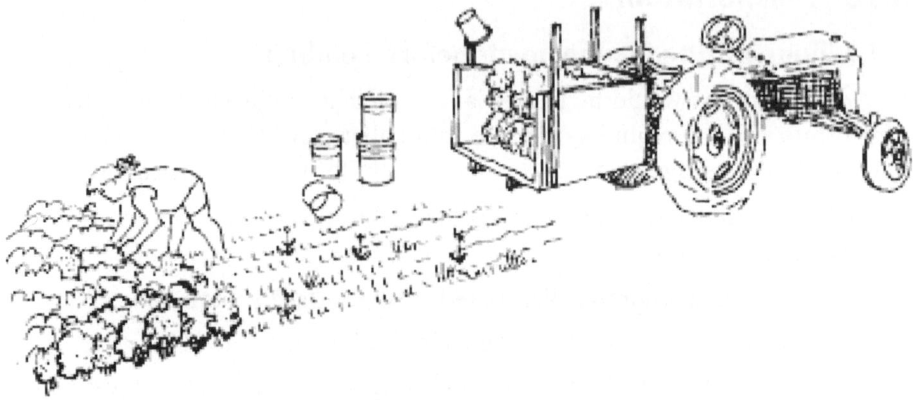

- ☆ A wider bed requires the harvester to bend over and reach in farther to harvest flowers from the center portion of the bed than would a narrow bed.
- ☆ A more awkward position will slow harvest and creates a greater distance to move individual flowers.
- ☆ The longer production beds are, the greater the distance flowers may have to be hand-carried to a trailer for transport.
- ☆ The transport system designed for a crop to be dried is harvested at the same time.
- ☆ A three-sided platform is mounted on the three-point hitch of a tractor.
- ☆ The flower crop is cut and placed directly into the bin with the tractor being backed down the bed as the harvest progresses.
- ☆ The harvest worker does not carry the cut stems anywhere, but merely turns around to place the crop into the bin.
- ☆ Only large quantities of the harvested crop at a time are moved any distance.

Chapter 4

Maintain Health and Safety at Work Place

4.1 Health Hazards at Work Place

4.1.1 Introduction

Health is a legal requirement. All small businesses such as flower shops/nurseries must have an occupational health program to prevent workplace injury and disease. Such programs must meet certain standards, and workers/floriculturist should exercise the due diligence in taking steps to meet those standards.

Health and Safety Involve the following Components

1. Hazard identification and risk control–Determine which hazards are present in the workplace and take steps to eliminate or minimize them.
2. Safe work procedures–Describe in writing how to carry out specific tasks safely.
3. Orientation, education, training, and supervision–Prepare workers for the job and make sure they continue to work safely. This is particularly important for new and young workers.
4. Safety inspections–Identify workplace hazards so that they can be eliminated or controlled.
5. Incident investigation–Find out why an accident or injury occurred so the causes can be corrected.
6. Health and safety meetings–Provide an opportunity for workers and supervisors to communicate any concerns about health and safety.

7. First aid–Determine what level of first aid is required for your workplace, and make sure everyone knows what to do if someone is injured on the job.
8. Records and statistics–Maintain documentation to help identify recurring problems, and ensure that hazardous conditions are corrected.

4.1.2 Risk at Workplace

Risk is a chance that someone could be harmed by hazards like working from ladders or working around electricity. A risk assessment is an important step in protecting people, business and complying with applicable laws.

Many tools/chemicals used by floriculturist/nursery workers are hazardous or dangerous. For example, fertilizers, solvents, and cleaners may cause conditions ranging from minor skin irritation to serious injury or death.

Floriculturist/Gardeners and workers are particularly exposed to certain hazards such as:

- ☆ Tiring or painful positions
- ☆ Carrying or moving heavy loads
- ☆ Standing or walking and repetitive hand or arm movements
- ☆ Work-related musculoskeletal problems
- ☆ Inhaling particulate matter
- ☆ Burns from cleaning chemicals or fertilizers and additives
- ☆ Cuts from blades
- ☆ Uneven or slippery surfaces
- ☆ Noise from equipment such as string trimmers

Floriculturists/gardeners live with risk and make decisions every day that affect their operations. Many factors that affect the decisions which gardeners/floriculturist make could not be predicted accurately such as:

- ☆ Change in weather conditions
- ☆ Drop in prices at the time of harvest
- ☆ Non-availability of hired labor during working times
- ☆ Breakage of machinery and equipment when required
- ☆ Change in government policies

Some more common hazards faced by floriculturist include:

- ☆ Operating equipment and power tools
- ☆ Pushing and lifting heavy equipment

- Using sharp tools such as pruning shears and picking knives
- Working with pesticides

4.1.3 Sources of Risk

The most common sources of risk in gardening/flowering can be in areas of:

- Production: Production risk occurs from the uncertainty of factors that affects the quantity and quality of flower production.
- Marketing: Marketing risk occurs due to the variability of product prices and uncertainty of future market prices.
- Financial: Financial risk occurs when money is borrowed to finance the flower business.
- Institutional: Institutional risk occurs due to unpredictable changes in the provision of services.
- Human: Human risk involves risk to flower business that is caused by human illness and personal situation of gardener/floriculturist family.

Based on floriculturist's attitude, risk can be divided among:

Risk-averse Floriculturist

- The risk-averse floriculturist avoids taking risks.
- They tend to be more cautious individuals with preferences for less risky sources of income.
- They sacrifice a certain amount of income to reduce the chance of low income and losses.
- They do not refuse to accept any risk at all.
- The risk-averse floriculturist seek to be compensated for the risk taken by receiving a higher return than would normally be obtained if there were no risks.

Risk-taker Floriculturist

- The risk-taker floriculturist is people who are open to more risky business options.
- They choose the alternatives which give some chance of higher outcome, even though they may have to accept a lower outcome.
- They are influenced by the return they could receive.

Risk-neutral Floriculturist

- Risk-neutral lies between the risk-averse and risk-taking positions.

4.1.4 Workplace Risk

The main causes of risk/accidents due to equipment while working at the site include:

- ☆ Transport (running over or vehicle over turnings);
- ☆ Falling from height (from fragile roofs, trees, *etc.*);
- ☆ Struck by moving or falling objects (bales, trees, *etc.*);
- ☆ Trapped by something collapsing or overturning;
- ☆ Livestock related accidents and fatalities;
- ☆ Asphyxiation/drowning.

Floriculturist can prevent most workplace injuries and illnesses if he/she identify workplace hazards and take steps to control them. Risk control involves eliminating the hazard entirely or, if that is not possible, minimizing the risks as much as possible. Ways to minimize the risks include the following:

- ☆ Select appropriate safety features when purchasing or replacing equipment.
- ☆ Modify work processes or equipment.
- ☆ Develop and implement safe work procedures for hazardous tasks.
- ☆ Ensure that workers use appropriate personal protective equipment and follow safe work procedures.

4.1.5 Improving Workplace Safety

Creating a safe workplace is an ongoing commitment. Floriculturist can make safe place by doing the following or ensuring that it is done:

- ☆ Providing regular safety training for supervisors/workers.
- ☆ Show an example and demonstrate safe working habits.
- ☆ Regularly check workers about safety procedures.
- ☆ Regularly check workers for using protective equipment/devices
- ☆ Encourage workers to report illnesses and injuries immediately.
- ☆ Encourage workers to report potential hazards.
- ☆ Keep workplace health and safety programs current.
- ☆ Stay aware of workplace hazards and how to handle them.
- ☆ Respond promptly to all health and safety concerns.

4.2 Practice General Safety and First Aid

4.2.1 Ensuring Safety

Most production nurseries and retail flower centers are small to medium-sized businesses. The workers of such small businesses don't automatically understand their occupational health and safety obligations and don't need written occupational health and safety procedures.

The way to prevent injury and ill health at the workplace and avoid breaching strict new occupational health and safety regulations is for all businesses to have effective health and safety policy and program in place which should cover clearly defined set of written safety roles and responsibilities.

Safe practices and safety conscious attitude by floriculturist/staff is the best protection against workplace accidents. To help in assessing the safety status of the flower business, floriculturists should answer the following questions, highlighting what's missing and determining where he/she needs improvement. By completing such questions, floriculturists can able to monitor his/her workplace health and safety performance which ensures his/her regular improvement.

4.2.2 Ways to Keep Employees/Customers Safe at the Workplace

Proper safety is important to any operation, as day-to-day activities of production/sales can take in focus on things like safety. To fight with combat that, floriculturist should make sure that he/she should regularly talk about safety practices and procedures with owners, managers, and workers. Following are the points which should be managed to handle safety procedure for employees/customers at the workplace:

1) Good Housekeeping Practices

- ☆ Keeping aisles and walkways clear and even/level
- ☆ Adequate clearance of exits, breaker boxes, emergency lighting, fire extinguishers, *etc.*
- ☆ In case of an emergency evacuation, helps people to see and identify safety equipment, prevents accidental slips/trips/falls, and prevents fires

2) Accident Investigations

- ☆ Identifying root causes
- ☆ Implementing corrective actions to prevent re-occurrence
- ☆ Gives employer documentation to record the information so it can be reviewed in the future if needed

3) Safety Training

- ☆ Promotes awareness
- ☆ Educates employees on safety with equipment, safety devices, chemical safety, fire prevention, personal protective equipment (gloves, safety glasses, hearing protection)

4) Internal Safety Inspections

- ✰ Employee/employer or team walks through the facility regularly with the purpose of identifying hazards.

5) OSHA Required Written Programs

- ✰ Allowing employers to put an organized plan in place
- ✰ Writing of safety programs line emergency action plan, Hazard Communication, Respiratory Protection Program, Lockout/Tagout, Confined Space, *etc.*

4.2.3 Reducing Risk of Hazards at the Workplace

There are many controls that floriculturist/employers can use to prevent hazards in the workplace. Firstly though, it is important to complete hazard identification and a risk assessment in consultation with staff workers for ensuring the selection of right control for hazards that are relevant in the workplace.

Common controls used in workplaces can be categorized as per hierarchy of controls such as:

- ✰ Eliminate hazard - install more power points to avoid cords on the floor, widen aisles
- ✰ Substitution - resurface floors with 'less hazardous materials'
- ✰ Isolation - restrict access to some work areas
- ✰ Engineering controls - improve lighting, mark walkways install drainage, use ramps instead of steps
- ✰ Administrative controls - ensure good housekeeping - clean up spills immediately, use signs for slippery or wet floors
- ✰ Personal Protective Equipment proper footwear

4.2.4 Providing Treatment Appropriate to Injuries with Recognized First Aid Techniques

While working in a shop or at the site, workers/labors involve in certain types of injuries. It is the duty of floriculturists to keep/arrange first aid tools at shop/site so that in case of any emergency, an immediate first aid could be given to the patients using first aid techniques.

First Aid Kits and Attendants

- ✰ Make every worker working in a shop aware of where the first aid kit is located and how to call the first aid attendant if required.
- ✰ Write the signs showing how to access the first aid.

- ☆ If a first aid attendant is required, make sure that the attendant holds a first-aid certificate of the level necessary for that workplace.
- ☆ If floriculturist needs first aid attendant, he/she should train the backup attendants. Floriculturists should also ensure that enough workers are trained for such responsibility to cover vacations and other absences.

Workplace Injuries

Common flower shop/site injuries could be bleeding, burns, eye injury, chemicals inside the eyes wounds and breathing, *etc.*

Control of Bleeding

- ☆ Direct pressure – use your hand(s).
- ☆ Elevate (raise) the limb
- ☆ Apply a pad and firm bandage.
- ☆ If necessary use cleans rags or clothing.

Burns

- ☆ Cool the burnt area with cool water for 10-15 minutes.
- ☆ If necessary, cover the burn with a clean dressing or plastic wrap before removing a person to medical aid.

Eye Injuries

Foreign Bodies in the Eye(s)

- ☆ Wash the eye(s) with eyewash or clean water.
- ☆ If the foreign body is stuck to the eye surface, do not attempt to remove it.
- ☆ Place a covering over the eye(s) and send for, or take, the person to medical aid.

Chemicals in the Eye(s)

- ☆ Wash the eye(s) with clean cool water for at least 15 minutes.
- ☆ Wash from near the nose outwards and always wash under the upper eyelid.
- ☆ Send for, or take, person to medical aid.

Minor Wounds

Abrasions, Small Cuts

1. Clean the wound with mild antiseptic or soap and water.
2. Cover lightly with a clean dressing.

Breathing

If a person is breathing but unconscious, turn them onto their side to prevent airway obstruction (tongue or vomit).

4.2.5 Handling/Storing of First Aid Equipment

Immediately and properly examining and treating injuries save life and are essential to reduce pain and help injured people make a quick recovery. All this is possible at shop/site through the first aid kit which should comprise of the following:

- ☆ Band-Aids
- ☆ Absorbent compress dressings
- ☆ Sterile gauze pads
- ☆ Adhesive cloth tape
- ☆ Gloves
- ☆ Roller bandages
- ☆ Antiseptic wipes
- ☆ Antibiotic ointment
- ☆ Scissors and tweezers
- ☆ Sterile saline
- ☆ Hydrocortisone ointment

It is important for floriculturists to check the first aid kit on a regular basis. Medications expire, so floriculturist needs to replace the old ones and should replenish the supplies which have been used earlier, so he/she should make sure that more common treatments like medications, bandages, *etc.* are always available. If possible use antiseptic solution to clean the scissors and injections.

4.2.6 Reporting First Aid Procedure at Workplace

Floriculturists/supervisors must provide information about first aid to their workers so that they should know what to do and whom to contact in case of any emergency. Information should be easy to understand, accessible and should take into account the language and literacy levels of the workers. Information may be given using verbal methods or visual methods which include:

- ☆ Location of first aid equipment and facilities
- ☆ Names and location of persons trained to administer first aid
- ☆ Procedures to be followed when first aid is required.

The information and instruction should be provided by floriculturists as part of workers' induction training when there are changes in the location of first aid facilities or names, contact details of first aiders.

It is the duty of the floriculturist or supervisor to check that medical person should keep all the details regarding the type of treatment, what he/she is giving along with details of first aid equipment that are dispensed. Floriculturists should record maximum information about the patient and type of injury, what the patient is suffering from, with old patient history. Such information is important when handing-over casualty to health care professional by covering:

- ☆ Name and contact number of casualty;
- ☆ Nature of injury;
- ☆ Time and date of the incident;
- ☆ Details of treatment given

4.3 Maintain Clean and Efficient Workplace

4.3.1 Introduction

Cleaning is a process of getting a clean, free and energetic environment where there is no risk of health hazards. Clean surrounding makes it easier to organize the workplace so that it is easy to access all sort of material which are required.

Cleanliness is an important quality of life and serves as a habit that is often considered just next to godliness. It is a habit that is not related to the wealth of a person, but which depicts the qualities of a person.

Cleanliness is nothing that should be forced on a person but should be properly practiced and be made a lifestyle. A clean and well-maintained person will always succeed in all difficulties if he/she applies it in every path he/she walks. It also helps to remain fit and healthy, mentally, physically and socially.

Cleanliness is required for the health and wellness of individuals and society. Health and cleanliness relate to each other and influence each other. In order to be in good health, one should practice hygiene. Hygiene is the practice of maintaining good health and preventing diseases through cleanliness. Cleanliness of body, food, and environment contribute to the wellbeing of individual health. It prevents disease as dirt usually dock on infected pathogens which cause the disease to human beings when they are exposed to them.

4.3.2 Cleanliness at Workplace

A clean workplace can be said to as a place where there is all-round protection from germs to health. It is not a place which looks clean but is a place which is actually germs free for the workers and customers. A clean and organized workplace environment will not just be beneficial for floriculturist/worker but also will help in getting more customers. Cleanliness helps in increasing productivity as dirty place and clean mind cannot work together.

Workplaces in many years have realized the importance of clean and healthy surrounding. There are many reasons for which, floriculturist should maintain cleanliness and hygiene at workplace. As workplace is where floriculturist/workers spend several hours during their working, so it is advisable that he/she should make the place clean, germs-free and healthy.

In workplace, the same concept applies as productivity is influenced by the environment, so clean environment will tend to increase productivity, as people will work with positive mindset that is free of stress. In health, cleanliness is important as hygienic practices contribute to good health of individuals. A dirty environment predisposes customer/staff to diseases, which affects their ability to perform activities and it results in less productivity.

4.3.3 Health and Safe Procedures in Workplace

There are certain health and safety procedures which should be followed by floriculturist/gardener/workers while working in workplace such as:

1. **Dry and Clean Slip Proof Floors/Pathways**

- ☆ Slips and falls are one of the common ways of enduring injuries, so to prevent injuries, the floors/pathways should be cleaned on regular basis.
- ☆ While planting seed, mud or water mixture creates slip, so soon after the planting of seeds, the working area should be clean.
- ☆ Slip can also occur due to wet contaminants such as water, oil or grease, or dry such as dust, metal, plastic bags or off-cuts.
- ☆ Preventing floor contaminants is one of the best things you can do to prevent slips.
- ☆ More serious slips or trips together with the resulting falls may result in:
  - ❖ Sprains or strains
  - ❖ Broken bones when trying to break the fall
  - ❖ A back injury due to the sudden and forceful impact during a fall
  - ❖ Burns if it occurs near hot surfaces or if the person is handling hot fluids
  - ❖ Cuts if it occurs near sharp objects.

2. **Unorganized and Messy Environment**

- ☆ The messy environment is the main cause of all problems.
- ☆ In this, there are more chances of threats going unnoticeable and the situation may simply turn the place into accident-prone area.
- ☆ An occupational hazard is a result of ignorance on part of workers or carelessness on part of workers.

- If workplace is messy and unorganized, it may be a door to many uninvited hazardous accidents, so a gardener should properly arrange and clean the working area from tools, chemicals, fertilizers *etc.*
- Proper piling of pots, fertilizer bags, and chemical boxes should be done in the store area, otherwise locating the items from improper piling will lead to stack falling upon workers and can make him/her injured.
- An unorganized store may also hamper the productivity rate of the worker as for locating particular item from the store without proper stocking will lead to time wastage of worker.

3. Use of Disinfectants in Workplace

- Disinfectants are very important to be used in garden, parks, public places as there are more risk of germs spreading around due to multiple uses of each and every item.
- In workplaces, there is frequent use of items from one worker to another, whether it is flu season or not, as germs are inevitable in nature.
- The disinfectants comparatively lessen the chance of germ produced illness or diseases.
- In order to cut the cost some workers opt for cheap disinfectant, which on longer run can affect the health and in turn, will lead to health issues and absentees.

4. Germ Controls to Fight Flu/Diseases

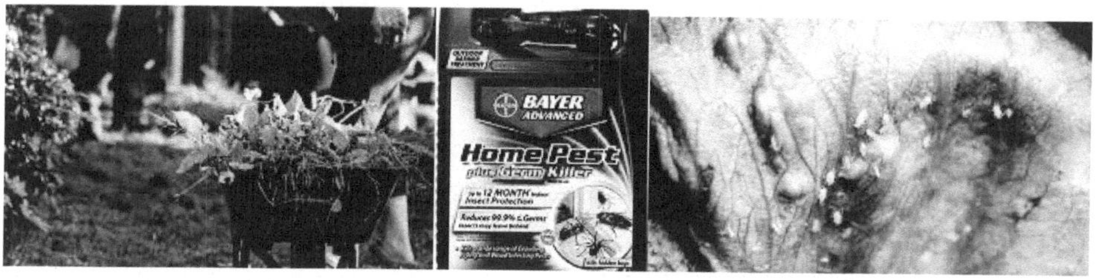

- ☆ Germs can easily disperse across the entire workplace if there is any compromise in the cleanliness.
- ☆ The sitting place or lunchroom can be called as germ house for workers, so it is important to clean and sanitize such areas frequently.
- ☆ Apart from flu, there is high chance of catching fatal disease or infection by shared germs in the workplace.
- ☆ There are many infectious germs that may spread due to unhygienic environment and lead to various disease such as fungal infection which can lead to irritation and skin infections.

5. Green and Clean Equipments/Products

- ☆ Eco-friendly products are best in nature and are not harmful to human health.
- ☆ Such products are friendly to nature.
- ☆ Not all cleaning products can be compared to one another as all such products have different standards.
- ☆ Some products are cheaper and harmful whereas some are costlier and trustworthy.
- ☆ It is very important to choose the correct chemical.
- ☆ While choosing chemicals, it is equally important to store these chemicals in correct places.

- There are more chances of such chemical to cause skin related allergies, which on longer run can affect the worker's attention span and confidence.

6. Air Filtering and Humidity

- Airborne diseases do not stay back in race of harming and spreading more and more diseases.
- One may not be able see them, but they can be equal threat in the way of the organization at a whole.
- A better ventilation system works as backbone against airborne diseases.
- Germs that can pass through respiratory infections can be tackled by ventilating system.
- It will not only enhance the productivity of workers but also maintain their good health.
- Thus, regular cleaning and timely changing of the required filters is very much advisable. Germs restrict the potential microbial growth or odor related problems in work environment.
- Humidity affect the germ level in workplace, so maintaining proper humidity level in store room or sitting area is advisable.

7. Proper Lightening

- Dirty light fixtures are indeed one reason to inefficient light supply.
- It can be a great hassle in way of workers.
- It is proven that adequate light is required in order to do daily task.
- No proper lightening can be an unsafe combination with workplace.
- Garden lawns, pathways, store rooms and equipment areas should be well lighted to prevent accidents and contribute to healthy and safe environment.

8. **Perfect Waste Disposal Plan**

   - ✰ Piled up trash will result in dirt all around the workplaces.
   - ✰ Piled trash is a place where pesticides can reside and multiply.

   - ✰ Pesticides can be a great threat and hazard to working environment.
   - ✰ Placing dust bins or waste bins throughout the workplaces is a way to gather waste in a single place.
   - ✰ Proper collection of waste from dustbins should be followed so that the waste will not spread anywhere in working area.

- ☆ Gardener/worker should implement reducing and recycling methods in workplace.

9. Maintaining Impression

- ☆ Workplace should be clean as apart from workers, there are customers, suppliers or sellers who used to roam in workplace.
- ☆ If the environment of workplace is messed up, unorganized and dirty, then there will be less possibility of good customers to visit and buy the product.
- ☆ Clean workplace will give better impression to the customers and will lead to good product sale.

4.3.4 Signs of Poor Workplace Cleanliness

The importance of housekeeping at work can be clearly felt when you see the surrounding neat and clean as you enter the workplace. Though every flower shop/ company comprises of housekeeping staff, it is not mandatory that they solely are responsible to maintain the cleanliness. Workers also need to understand the importance of cleanliness in the workplace and practice a few workplace hygiene tips so as to maintain a happy and clean shop environment.

There are many signs which reflect poor cleanliness at the workplace such as:

- ☆ Cluttered and poorly arranged work areas
- ☆ Untidy or dangerous storage of materials
- ☆ Dusty and dirty floors and work surfaces
- ☆ Items lying on work paths which are in excess or no longer needed
- ☆ Blocked or cluttered work area or path
- ☆ Tools and equipment left in work areas instead of stored properly
- ☆ Broken pots, containers, and damaged materials

- Overflowing waste bins and containers
- Spills and chemical leaks

4.3.5 Cleaning and Hygiene Rules for Workers

- No cluttering of material used
- Always use gloves and proper hand sanitizer
- Make use of rubbish bin and proper damping policy
- Removing of unwanted plants, grass from flowers
- Usage of proper working tools
- Keeping a proper stock of chemicals and other materials properly
- Have a regular discussion on health and safety matters
- Disposal of boxes and packaging covers should be done immediately

4.4 Render Appropriate Emergency Procedure

4.4.1 Introduction

An emergency procedure is a plan of actions that need to be done in certain order or manner, in response to a specific class of foreseeable emergency, which poses an immediate risk to health, life, property, or environment.

There are many emergencies that require urgent intervention to prevent a worsening of the situation, while in some situations, mitigation may not be possible and agencies will offer only palliative care for the aftermath. In such a case, the emergency plan should allow for certain possibilities.

An emergency procedure should provide an appropriate and proportionate response to certain situations and should identify the responsibilities, actions, and resources required to deal with an emergency situation.

Once everything has been drafted, a procedure should be followed by the consultative period with those who involved/affected by an emergency, and a program should be set out for testing, training and periodic review.

4.4.2 Workplace Emergency

Emergencies at the workplace include:

- Floods
- Fires
- Toxic gas releases
- Chemical spills
- Radiological accidents
- Explosions

- ☆ Civil disturbances
- ☆ Workplace violence

4.4.3 Emergency Response Plans

Flower shops/nurseries should be prepared to respond to emergencies such as fires, chemical spills, or natural disasters. If an emergency occurs, there will be a need to make quick decisions that will minimize injuries and damage. Such decisions are easier if you have already developed an emergency plan.

Follow these guidelines to develop and implement a plan:

- ☆ List all possible events such as serious injuries, fires, explosions, or natural disasters.
- ☆ Identify the major consequences associated with event-related to casualties, equipment damage, or facility damage.
- ☆ Determine the necessary measures to deal with consequences related to first aid, notification of medical authorities, rescue, firefighting, or equipment evacuation.
- ☆ Determine the type of resources required such as medical supplies or rescue equipment.
- ☆ Store emergency equipment where it will be accessible in case of emergency.
- ☆ Ensure that workers are trained in emergency procedures and shown where equipment is stored.
- ☆ Floriculturist to hold periodic drills at least annually to ensure that workers will be ready to act in case of any emergency.
- ☆ Floriculturist communicate the plan to everyone whoever is working in the workplace.

4.4.4 Role of Floriculturist during Emergency/Evacuation

- ☆ Assessing the situation to determine whether an emergency exit requiring activation of the emergency procedures;
- ☆ Supervising all efforts in the workplace, including evacuating personnel;
- ☆ Coordinating with outside emergency services such as medical aid and local fire departments and ensure their presence if necessary;
- ☆ Directing the shutdown of workplace operations when required.

4.4.5 Fire and Emergency Procedures

It is the duty of floriculturist/supervisor to know and understand the following procedures in case of fire in workplace:

1. Raising the alarm and evacuating people from the work area
2. Calling emergency services
3. Calling the manager/in-charge
4. In case of the small fire, use a fire extinguisher to extinguish the fire
5. If the fire is near the fuel tank, do not extinguish the fire, but retreat to a safe distance
6. Be prepared to direct the fire service to the fire area

At the workplace, emergency procedures depend on worksite which includes:

- ☆ Emergency shutdown
- ☆ Evacuation
- ☆ First aid
- ☆ Fire fighting

1. Emergency Shutdown of Equipment

- ☆ If there is a fire, emergency or accident, in such case floriculturists should stop using the equipment.

2. Emergency Evacuation Procedures

- ☆ On hearing an evacuation alarm, or on the instruction of emergency control personnel, immediately cease all activity and safeguard personal valuables.
- ☆ Help any person in immediate danger, but only if required to do so.
- ☆ Safeguard any activity/process that results in hazardous or suffer damage if left unattended as a consequence of evacuation.

- Act as per directions given by emergency control personnel and evacuate the work area immediately.
- Assist with general evacuation if directed to do so by emergency control personnel.
- Assist with evacuation of disabled occupants.
- In a fire, do not try to work in fields.
- Move calmly to the required evacuation area and do not leave the evacuation area until everything has been cleared.
- Follow the instructions of relevant emergency services personnel.

3. First Aid

- First Aid is the quick care given to an injured or ill person.
- If somebody needs first aid, floriculturist must tell their supervisor or manager.

4. Fire Fighting Equipment

- Fire fighting equipment on-site could be anything from small fire extinguishers to large water cannons.
- Different fire fighting equipment should be used for different types of fire.
- Always check the equipment for information on what type of fire it can be used on.

Steps for Using a Fire Extinguisher

1. Evacuate the area.
2. Isolate the area.
3. Call emergency services or other designated on-site procedures.
4. If it is safe to do so, use an extinguisher to attempt to control the fire.

4.4.6 Spill Procedures

It is the duty of floriculturist/supervisor to know and understand the following procedure in case of occurrence of spill:

- If the spill is from the hose or tap, shut the isolation valve
- Warn people in the area of the spill and evacuate, if necessary
- Remove sources of ignition if flammable substances are there
- Weigh up the spill by only response if think it is safe to do
- Refer to the safety data sheet or call on an approved handler or other specialists for advice

- If necessary, call emergency services and take advice from the local authority
- Put on safety equipment (boots, gloves, eye protection, *etc.*)
- Contain the spill if it is safe to do so using a drip tray or oversize container or spill kit to soak the substance
- Dispose of waste safely

4.4.7 Using an Emergency Action Plan

An emergency action plan covers required actions that workers/labors must take to ensure safety from fire and other emergencies.

Emergency action plan must include:

- A required method for reporting fires and other emergencies;
- An evacuation policy and procedure;
- Emergency escape procedures and route assignments like floor plans, workplace maps, and safe or refuge areas;
- Details of individuals both within and outside the workplace to contact for additional information/explanation of duties and responsibilities under the emergency plan;
- Procedures for workers who remain to perform or shut down critical workplace operations, operate fire extinguishers or performing other essential services which cannot be shut down for every emergency alarm before evacuating;
- Rescue and medical duties for any workers designated to perform them.

Index

I

K

L

M

N

O

P

R

S

T

Flowers Gardening Calender

January

- ☆ Winter Jasmine
- ☆ Crocus
- ☆ Witch Hazel

February

- ☆ Galanthus-Snowdrop
- ☆ Iris Danfordiate
- ☆ Hellebore

March

- ☆ Forsythia
- ☆ Daphne

April

- ☆ Daffodils
- ☆ Trillium
- ☆ Spirea

May

- ☆ Lilac
- ☆ Peonies
- ☆ Primula

June

- ☆ Hibiscus
- ☆ Rose
- ☆ Sunflower

July

- ☆ Roses
- ☆ Marigold
- ☆ Sunflower
- ☆ Lotus
- ☆ Lilies

August

- ☆ Hibiscus
- ☆ Jasmine

September

- ☆ Hyacinthus
- ☆ Snapdragon
- ☆ Daffodil

October

- ☆ Calla Lilies
- ☆ Sunflower
- ☆ Delphinium

November

- ☆ Rose
- ☆ Dahlia
- ☆ Tulip

December

- ☆ Petunia
- ☆ Cineria
- ☆ Marigold
- ☆ Violet
- ☆ Gerbera

Seasonwise Flowering Plants

Winter Season

- Calendula
- Petunia
- Pansy
- Dahlia

Summer Season

- French Marigold
- Sunflower
- Gaillardia

Flowers Growing Throughout the Year

- Rose
- Bougainvillea
- Hibiscus
- Ixora
- Crossandra

9 789354 614675

Printed at Repro India Ltd.